W0043906

Immunoassays for Food-poisoning Bacteria and Bacterial Toxins

FOOD SAFETY SERIES

General Series Editors
J. Edelman *London, UK*
S. Miller *Texas, USA*

Series Editor — Microbiology
T. Roberts *Reading, UK*

Editorial Board
D. Conning *London, UK*
D. Georgala *Reading, UK*
J. Houtvast *Wageningen, The Netherlands*
C. Mercier *Paris, France*
E. Widdowson *Cambridge, UK*

Forthcoming Titles
Food Reactions
M. Lessof

The *Staphylococci* and their Toxins
M. Bergdoll

The *Aeromonas* Group as a Foodborne Pathogen
S. Palumbo and F. Busta

Food Preservation
G. Gould

Fibre and Complex Carbohydrates
I. Johnson and D. Southgate

Immunoassays for Food-poisoning Bacteria and Bacterial Toxins

G. M. Wyatt

with contributions from
H. A. Lee and M. R. A. Morgan
AFRC Institute of Food Research, Norwich, UK

Springer-Science+Business Media, B.V.

First edition 1992
© 1992 G. M. Wyatt, H. A. Lee and M. R. A. Morgan
Originally published by Chapman & Hall in 1992

Typeset in 12/13pt Garamond 3 by Columns of Reading Ltd
Softcover reprint of the hardcover 1st edition 1992

ISBN 978-1-4613-5826-8   ISBN 978-1-4615-2001-6 (eBook)

DOI 10.1007/978-1-4615-2001-6

Apart from any fair dealing for the purposes of research or private study, or
criticism or review, as permitted under the UK Copyright Designs and
Patents Act, 1988, this publication may not be reproduced, stored, or
transmitted, in any form or by any means, without the prior permission in
writing of the publishers, or in the case of reprographic reproduction only in
accordance with the terms of licences issued by the Copyright Licensing
Agency in the UK, or in accordance with the terms of licences issued by the
appropriate Reproduction Rights Organization outside the UK. Enquiries
concerning reproduction outside the terms stated here should be sent to the
publishers at the London address printed on this page.

The publisher makes no representation, express or implied, with regard to
the accuracy of the information contained in this book and cannot accept any
legal responsibility or liability for any errors or omissions that may be made.

A catalogue record for this book is available from the British Library

Library of Congress Cataloging-in-Publication data available

Contents

List of Figures

Series Introduction

Consumer safety has become a central issue of the food supply system in most countries. It encompasses a large number of interacting scientific and technological matters, such as agricultural practice, microbiology, chemistry, food technology, processing, handling and packaging. The techniques used in understanding and controlling contaminants and toxicity range from the most sophisticated scientific laboratory methods, through industrial engineering science to simple logical rules implemented in the kitchen.

The problems of food safety, however, spread far beyond those directly occupied in food production. Public interest and concern has become acute in recent years, alerting a wide spectrum of specialists in research, education and public affairs.

This series aims to present timely volumes covering all aspects of the subject. They will be up-to-date, specialist reviews written by acknowledged experts in their fields of research to express each author's own viewpoint. The readership is intended to be wide and international, and the style to be comprehensible to non-specialists, albeit professionals.

The series will be of interest to food scientists and technologists working in industry, universities, polytechnics and government institutes; legislators and regulators concerned with the food supply; and specialists in agriculture, engineering, health care and consumer affairs.

One of the most difficult situations to control is the contamination of food by small numbers of pathogenic micro-organisms before they multiply to give the large populations causing food poisoning when eaten.

A rapid detection method would be of immense value to food producers and retailers, to public analysts and legislators as well as benefiting the public at large at a time when the reported incidence of food poisoning appears to be rising year by year. This book, by Drs Wyatt, Lee and Morgan, describes the current state of research into the powerful and elegant techniques of immunoassay as a means of dealing with this microbiological problem.

J. Edelman

Acknowledgement

The authors wish to acknowledge the Photography Section at the Institute of Food Research, Norwich for excellent preparation of the figures, and Mrs Janet Arnett for careful typing of the manuscript.

Preface

This book is intended to be a compilation or distillation of current knowledge of the subject, and most of the material in it is within the experience of the authors. Where necessary, specific references to the literature have been made; otherwise a bibliography is given at the end of the appropriate chapter.

In order not to interrupt the flow of the text, detailed methodology has been consigned to a series of appendices.

CHAPTER

—1—

Introduction

At the end of the 1980s, public awareness of the possible presence of food-poisoning bacteria in the food supply of the developed countries greatly increased. Examples receiving much publicity in the United Kingdom included outbreaks of salmonellosis originating from unprocessed eggs and the emergence in the public eye of *Listeria monocytogenes* as a potential pathogen (Figure 1.1).

In contrast to many other aspects of life in the Western World, where considerable improvements had been made in the quality of life, the contemporary diet came to be seen as a potential cause of morbidity and mortality. Although, in the home and workplace there had been a great deal of newly introduced technology, the methods of food analysis for detecting these pathogens were, in the 1980s, still firmly rooted in the 1880s, relying on ideas developed by the great pioneers of bacteriology such as Koch, Pasteur and Lister.

Clearly, in the public perception these methods were failing to detect pathogens in food or, more likely, were never actually being applied by the food producer, largely, it would seem, due to the cumbersome, labour-intensive and time-consuming techniques involved. However, in 1989, in response to public concern, the United Kingdom launched an investigation into the microbiological safety of food (HMSO, 1990, 1991); this was followed by passage of a Food Safety Bill through Parliament and this, together with other regulations such as the Poultry Orders, imposed additional responsibility on food

Figure 1.1
Food poisoning hits the headlines

producers to ensure that food destined for human consumption is not 'injurious to health'. The incentive for the considerable increase in testing of foods required by this legislation is blunted by the problems of traditional technology; if foods are to be certified as pathogen free (an impossible ideal?), then several days of expensive cold storage would be required while testing proceeded. Clearly, methods that are both faster and more convenient are required if producers are to fulfil their obligations. What happens to the food in the hands of the consumer is, of course, beyond the control of the producer and is largely a matter of public education.

Many approaches to the development of rapid methods have been taken, but for specific detection of pathogens the leading technology in the field is based on the use of antibodies in immunoassays. Relatively rapid methods for the total count of bacteria in a food sample have been available for several years, such as the direct epifluorescent filter technique (DEFT) and methods based on changes in electrical conductance or impedance in a culture medium during growth of bacteria. However, these methods are unable to discriminate easily between closely related genera of bacteria and are thus unsuitable for specific pathogen detection.

Immunoassays, in addition to their specificity and sensitivity, have another great advantage in that many formats are available. It should be stated here that, in the present context, the term 'immunoassay' might be replaced by 'immunotechnique' because some of the concepts to be covered in this book are not assays in the strict sense. The most widely used format at present in the non-clinical area is the enzyme-linked immunosorbent assay (ELISA). This can be based on a 96-well microtitration plate, the flexibility of which allows for analyses that are completely manual, completely automated or any stage in between. Enzyme immunoassays were developed from radio-immunoassays originally described by Yalow and Berson (1959) in which isotopes were used for end-point detection. Radio-isotopes present problems of safe disposal and, additionally, are not seen as acceptable in a food laboratory. An alternative, non-isotopic, end-point detection system was required and, in 1971, the ELISA, in which one component is immobilized and enzymically mediated colour reactions act as the end-point, was introduced (Engvall and Perlmann, 1971; van Weeman and Schuurs, 1971). A further important development came with the introduction of monoclonal antibodies (Kohler and Milstein, 1975) which have much greater potential specificity and which could theoretically be produced in almost unlimited quantities *in vitro*. ELISAs still took time to become accepted in the food analysis laboratory, where traditional methodology seems to have greater inertia than in the clinical laboratory. However, commercially available ELISAs for food-poisoning bacteria or their toxins do now seem to have been accepted and are becoming more widely used: some have received official approval as analytical methods. Other test systems based on the use of antibodies are emerging.

It is hoped that this book will provide the reader with at least two things: first, sufficient background to understand the theory behind any commercial assays used in their laboratory and to enable these to be adapted if necessary and validated for a particular use; secondly, to contribute, perhaps, to expansion of the field by developing assays for analytes not currently available. The authors have gained much experience in this area and are eager to pass it on!

CHAPTER
—2—

Food-borne Bacteria

Distinction has been made in the past between 'true' food-poisoning organisms, i.e. those that grow in food (with or without toxin production), and those that are merely transmitted by food. In either case their presence in food is a potential hazard and, thus, from both the consumer's and the analyst's standpoint they should be treated equally. 'Food-borne bacteria capable of causing disease' might be a better general term. However, for practical purposes, only those organisms that cause gastrointestinal symptoms are usually considered, excluding the occasional known transmission of brucellosis, tuberculosis, diphtheria and anthrax via food; *Listeria monocytogenes* and *Clostridium botulinum* should be regarded as exceptions to this view because their effects are not primarily in the gut. Viruses, fungi and parasites, although undoubtedly of importance, will not be considered.

It is useful to divide the food-borne bacteria into those causing infections of the gastrointestinal tract, usually with production of colonization factors and toxins (largely Gram-negative organisms), and those (largely Gram-positive organisms) producing toxins during growth in the food, ingestion of which gives rise to symptoms. In order to reach their site of action, these latter toxins would have to be stable in the acidic environment of the stomach, and to the action of proteases present in the small intestine. Components of food may, in some circumstances, protect toxins from degradation by these factors.

Enterotoxins, i.e. substances giving rise to symptoms in the gut, can be either exotoxins (secreted by live bacteria) or endotoxins (released on lysis of the bacterial cell). The latter term is also often used nowadays to refer to the surface lipopolysaccharide chains of Gram-negative bacteria (see Section 2.3.2).

—— 2.1 ——
INTOXICATIONS BY PRE-FORMED TOXIN

2.1.1 Staphylococcus aureus
On a world-wide basis *Staphylococcus aureus* is probably the most common cause of bacterial food poisoning and this is unfortunate because, in developed countries at least, it ought not to occur at all. Two main reservoirs of the organism exist: first, the organism is present on the skin and in the hair and nasal passages of a large number of healthy individuals, i.e. it is part of the normal flora of many food handlers. Good hygiene should, in fact, prevent the organism from reaching those foods that are particularly suited for growth and formation of toxins, e.g. cooked ham and dairy cream. The organism does not normally grow in uncooked meats; contamination of these foods is entirely post-processing and thus preventable. The second major reservoir is the udder of bovines, where it is involved as the causative agent of mastitis. Again, growth in dairy products should be prevented by pasteurization and refrigeration of the milk and its products.

The enterotoxins of *Staph. aureus* are simple, water-soluble proteins with molecular weights between 28 000 and 35 000. At least seven serologically distinct toxins are known, identified as A to F. Enterotoxins are not readily destroyed by heating, and cases are on record where small amounts of active toxin have survived canning. The enterotoxin probably acts on the lining of the abdomen and the onset of symptoms is characteristically rapid, normally within 6 hours of ingestion of the food. Nausea, vomiting and diarrhoea are usual, with collapse of the sufferer in severe cases.

The usual level of enterotoxin found in foods is about 0.5–10 μg/ 100 g food, so the limit of detection required for an assay needs to be below this value, preferably in the region of 0.1–0.2 μg/100 g.

2.1.2 Bacillus cereus

Unlike *Staph. aureus*, this organism is widespread in the environment and raw foods are frequently contaminated. Because it is a spore-forming species, cooking temperatures need to be considerably in excess of 100°C in order to kill the organism. Interestingly, the method of cooking and handling may select for the type of *B. cereus* poisoning which subsequently results from abused food; spores of serotype 1 are reported to be more heat resistant than the spores of other serotypes, and these strains are the most commonly isolated in one of the two variations of *B. cereus* intoxication – the vomiting-type poisoning. This syndrome has been associated especially with fried or boiled rice of the type used in some ethnic cooking, in particular involving Chinese restaurants. The practice of retaining portions of boiled or fried rice for up to 3 days without refrigeration (because refrigeration causes rice grains to stick together) allows growth of *B. cereus*, and the sequence of heating and reheating may enhance germination of spores by heat activation. Counts of *B. cereus* of greater than 10^9 cells/g rice have been recorded. The incubation period for this form of *B. cereus* poisoning is similar to that of *Staph. aureus* intoxication and, whilst vomiting is the principal feature, occasional diarrhoea is reported. The emetic agent is a heat-stable low-molecular-weight (<5000) peptide.

The other variation of *B. cereus* intoxication is the diarrhoeal syndrome. This is particularly associated with meat and meat products, vegetables and desserts made with corn starch, and has a longer incubation time, typically 8–16 h. The usual number of *B. cereus* organisms present in the implicated foods seems to be 10^7–10^8/g, and many serotypes are involved. The enterotoxin is a protein with a molecular weight of 55 000–60 000 and seems to be markedly heat labile.

2.1.3 Clostridium botulinum

The high fatality rate in cases of *Clostridium botulinum* intoxication has stimulated an enormous amount of effort to prevent the occurrence of poisoning; botulism is a rare form of food poisoning in the UK. Although it is a spore-forming organism similar to *B. cereus*, *Cl. botulinum* is anaerobic and requires a lowered redox potential for growth. There are, however, situations in which this may occur in foods, in particular where oxygen has been boiled off by cooking, or in vacuum packs. Canned foods offer a particularly favourable environment for growth, and the history of the development of retorting conditions used in canning has centred on elimination of every viable *Cl. botulinum* spore. For non-canned foods, growth of the organism can be prevented by combinations of acidity and curing salts such as sodium chloride and sodium nitrite. As with *B. cereus*, the spores are widely distributed in nature and it is likely that many raw foods will be contaminated.

The toxins are proteins of molecular weights of approximately 150 000, consisting of two linked principal chains of molecular weight 50 000 and 100 000, often associated with a non-toxic component to form higher-molecular-weight complexes. The proteins are released on lysis of the cells and then appear to be activated in the gut by proteases and absorbed into the blood stream. In at least some strains of the organism toxin production is mediated by bacteriophage infection and cultures freed of phage are non-toxigenic.

The toxins are thought to block the release of acetylcholine from nerve endings and therefore symptoms are of a neurological type, such as dizziness, blurred vision and muscle weakness. Death can occur due to respiratory failure.

—— 2.2 ——
INFECTIONS

The term 'infections' is used here in a rather loose sense to cover both those organisms that multiply in the lumen of the gastrointestinal tract and those that colonize or invade the intestinal mucosa. A very complex anaerobic flora exists in the intestine, with numbers of bacteria reaching more than 10^{11}/g of contents at the lower end of the large intestine. In higher regions of the intestine (duodenum, ileum and jejunum) the flora is considerably more sparse due to the flow of acidic gastric juice from the stomach. If an invading organism is to establish itself in the gut, then it must be able to compete successfully against the normal flora and adapt to the gut environment.

2.2.1 Salmonella
Salmonellae are high-profile organisms and probably constitute the food-poisoning bacterium best known to the general public in the West. In the UK, the officially recorded food-poisoning statistics show more cases due to salmonellae than all other organisms combined. (The

number of gastrointestinal infections recorded due to campylobacters exceeds the total due to salmonellae, but the species of campylobacter that cause enteritis are not yet officially recorded by the Public Health Laboratory Service (PHLS) as being food borne; this situation may well change in the future.)

Strains of salmonellae are divided into two main groups, the largest being the general non-host-adapted serotypes which cause simple gastrointestinal disturbance. The second grouping consists of the host-adapted serotypes which are capable of systemic invasion of the host: in humans these are *Sal. typhi* and *Sal. paratyphi*, the causative agents of the typhoid fevers. Serotypes host adapted to other animals and causing similar infections in those particular animals (e.g. *Sal. dublin* in cattle) are still capable of causing food poisoning in humans. Although around 2000 serotypes of salmonellae are known, most cases of food poisoning are caused by a much more restricted range of strains; in the UK in 1990, for example, *Sal. enteritidis* and *Sal. typhimurium* together accounted for more than three-quarters of all reported cases (PHLS, 1990).

Salmonellae multiply in the lumen of the small intestine causing an inflammatory reaction in the ileum. Diarrhoea, vomiting and abdominal pain occur after an incubation period of 12–36 h. Reports of the infective dose vary widely, but it must have been less than 100 bacteria in several recorded outbreaks, in particular where food components may have protected the bacterial cells from stomach acid. The very young and very old are more susceptible, and deaths are not infrequent in these groups. The involvement of toxins has not been proved, but the general Gram-negative lipopolysaccharide endotoxin may be implicated in pathogenicity.

2.2.2 Escherichia coli

As with *Staph. aureus*, *Escherichia coli* is another organism that is readily isolated from food handlers and poor hygiene must play a large part in dissemination. Many *E. coli* serotypes form a normal but minor constituent of the human intestinal flora without causing any harm, but a wide variety of potentially pathogenic strains exists and raw foods of animal origin are frequently contaminated. In the less developed countries, diarrhoea caused by *E. coli* is responsible for many deaths of babies and young children. The organism is spread by contaminated supplies of water and food.

Three main groups of pathogenic *E. coli* are now recognized. The first comprises the enteropathogenic strains (EPEC) which cause infantile gastroenteritis and are particularly associated with group care of the young, especially when they are already weakened by other conditions. High infective doses are required and these can be provided by rapid growth of the organism in food; further proliferation takes place in the small bowel but no clear role of toxins has been shown.

Enterotoxigenic *E. coli* (ETEC) serotypes are responsible for 'traveller's diarrhoea', especially during visits to warmer countries, and individual susceptibility to the infective strain is important. A combination of colonization factors, allowing adherence to intestinal mucosa, and enterotoxin production makes these serotypes pathogenic. Two toxins are produced: a heat-labile toxin (LT) which is a protein of molecular weight 86 000 related to cholera toxin, and a heat-stable (ST) polypeptide of molecular weight 5000.

A notable subgroup of ETEC strains is emerging as a major pathogen in North America and to a lesser extent in the UK. These are the Vero-cytotoxin producing *E. coli* (VTEC) which produce at least two toxins that are cytotoxic to Vero cells in culture. The toxins are both formed of two protein subunits, with molecular weights of approximately 33 000 and 8000. Most VTEC strains belong to serotype 0157 and the symptoms range from mild diarrhoea to a very severe bloody diarrhoea (haemorrhagic colitis); occasionally, renal failure occurs (haemolytic ureamic syndrome). It is believed that, in the USA and Canada, cattle are the reservoir of infection with meats and unpasteurized milk as the vehicles. A call has been made (Brook and Bannister, 1991) for better diagnostic methods for this group of organisms.

The third grouping is enteroinvasive *E. coli* (EIEC) serotypes which produce an illness very similar to dysentery; in fact, some EIEC serotypes are related antigenically to shigellae. Outbreaks of disease due to EIEC are uncommon, but cheese was involved in one incident in the USA.

The normal association of non-pathogenic *E. coli* strains with the enteric tract of humans and animals has led to the presence of these organisms in foods being taken as an indication of the possible presence of pathogens – or, at the very least, a lack of hygiene in handling the food is suggested.

2.2.3 Clostridium perfringens

As a species, this organism produces a range of toxic metabolites, including the potent necrotic toxin responsible for gangrene of infected wounds. However, involvement in food poisoning is mostly limited to *Clostridium perfringens* type A. The problems associated with an anaerobic spore-bearing organism were discussed under *Cl. botulinum* and apply in principle to *Cl. perfringens*. It can multiply extremely rapidly in the right conditions, and survival of spores in cooked meat dishes, which are subsequently inadequately cooled, commonly leads to the development of a large population of bacteria in the food. After ingestion, these bacteria sporulate in the gut and toxin production is associated with this phase of the organism's life cycle. Diarrhoea and abdominal pain follow 8–20 h after ingestion of the food.

The toxin is a protein with a molecular weight of about 35 000; it causes fluid accumulation in the small intestine but the exact mode of action is not known. A large amount of toxin (milligram quantities) is required to produce symptoms so, although it can be produced in foods, the amounts required are unlikely to be sufficient for direct intoxication after ingestion; additionally, the toxin is not resistant to acid or proteolytic enzymes.

2.2.4 Vibrios

Organisms of the genus *Vibrio* are essentially aquatic (in particular marine) bacteria and for optimum growth demonstrate a wide range of requirements for NaCl. Infections due to *Vibrio cholerae* probably produce many more mild cases of diarrhoeal disease than severe ones. This is especially true of the strain responsible for the current pandemic, the symptoms of which can be difficult to distinguish clinically from some *E. coli* infections. However, severe cholera is still a life-threatening infection unless treated; enormous quantities of fluid and salts are lost by diarrhoea and vomiting, and this can quickly become debilitating. The organism is markedly acid labile and so some temporary rise in stomach pH, probably due to food, seems to be necessary for it to reach the small intestine; here it multiplies and releases a protein exotoxin. The toxin has a molecular weight of about 84 000 and consists of two components, one of which appears to be responsible for interaction with cell membranes.

Contamination of food with *V. cholerae* is due to poor hygiene and usually originates from river and other poor-quality water supplies; as such, it is largely a disease of less developed countries where it can spread quickly. The first South American epidemic this century started in Peru in January 1991 and swiftly reached neighbouring countries.

In contrast, the other food-poisoning *Vibrio* species – *V. parahaemolyticus* – is the major cause of poisoning in one of the world's most developed countries – Japan – which still retains a tradition of eating raw fish and other seafoods. Profuse diarrhoea follows about 18 h after infection with this organism and severe abdominal pain is usual. Isolates of *V. parahaemolyticus* from patients almost always produce a haemolysin (which lyses red blood cells) but the role of this in pathogenesis is not clear. Isolates from implicated foods and marine environments are, conversely, almost always haemolysin negative, so an enrichment or conversion of some kind must take place in the gut.

2.2.5 Shigellae

The four species of *Shigella* currently recognized (*Sh. sonnei*, *Sh. dysenteriae*, *Sh. flexneri* and *Sh. boydii*) are closely related to *E. coli* and it has been suggested that they should be placed in the same genus. They are host adapted to humans and the presence in food results from faecal contamination due to poor hygiene in food handlers. Direct contact is also a route of infection because the infective dose of many strains is very low. The organism persists in food largely without multiplication and would probably not be a problem except for the low infective dose. Unlike most of the organisms considered here, the site of infection of shigellae is in the large intestine where it invades the colonic epithelial cells and multiplies. Necrosis and ulceration follow and bloody diarrhoea results. Formation of enterotoxin is of less importance in pathogenesis than the invasive ability.

2.2.6 Campylobacters

As mentioned in the section on salmonellae, campylobacters are reported as the major cause of gastrointestinal illness in the UK. The view that it is mainly food borne, once unconsidered, is now becoming accepted, although the organism is largely unheard of by the general public. Confirmed food-borne outbreaks have involved unpasteurized milk and poultry and it is likely that, as with salmonellae, poultry are a frequent vehicle of transmission. Surveys of retail poultry carcases have demonstrated widespread contamination. As for salmonellae, the natural habitat of campylobacters seems to be in the intestinal tract of humans and animals. The species responsible for food poisoning are *Camp. jejuni* and *Camp. coli* and the symptoms include abdominal pain and severe diarrhoea, often leading to prostration; the illness can last up to 2 weeks. Occurrence of infections has a very marked seasonal peak in the summer, the reason for which is not clear.

It is possible that, in many cases in the past, the organisms have been overlooked owing to their growth needs. *Camp. jejuni* and *Camp. coli* will not normally grow either aerobically or in strictly anaerobic conditions. There is a requirement for a microaerophilic atmosphere and it is usual now to provide 5% oxygen, 10% carbon dioxide and 85% nitrogen for optimal growth in the laboratory. They are markedly labile organisms, being sensitive to environmental conditions such as chilling and drying. Growth is not usual in foods and the organisms die out readily. However, the infective dose has been shown to be as low as 500 bacterial cells. These characteristics have probably been responsible for the pattern of campylobacter enteritis in humans – cases, although large in total number, tend to be individual rather than occurring in outbreaks such as incidents of salmonella poisoning, where the organism does multiply in foods.

2.2.7 Listeria monocytogenes

If salmonellae have long been known to the general public as a cause of food poisoning, then it has recently been joined in its position as a high-profile organism by *L. monocytogenes*, which has become a 'killer bug' to the UK popular press. As a cause of abortion, or listeriosis of the newborn, the emotional reaction is understandable although the actual incidence of proven cases is small.

L. monocytogenes is widespread in the environment and has been isolated from many sources, including numerous foodstuffs. It is resistant to many environmental stresses and can remain viable for long periods. Confirmed cases of food transmission have particularly involved unpasteurized milk and milk products such as soft cheeses. The ability of *L. monocytogenes* to grow relatively fast at refrigeration temperatures (many strains will grow at 2°C) means that normal cooling of food does not guarantee freedom from the organism; indeed, some enrichment at the expense of the other flora may take place and this feature has been used as an isolation technique in the laboratory.

The symptoms of listeriosis are not gastrointestinal, but appear as meningitis, septicaemia and many other systemic problems. Healthy adults are not normally infected, but any compromise in immune status, such as occurs in pregnancy, malignant disease or extremities of age, can render an individual susceptible. Both *L. monocytogenes* and *L. ivanovii*, which is regarded as an occasional human pathogen, produce a haemolysin but the role of the lysin in pathogenesis is not yet clear.

2.2.8 Other organisms

Many other organisms, especially those in the families Enterobacteriaceae and Vibrionaceae, can be regarded as possible food-poisoning bacteria. The potential of some of these is only just being realized and the term 'emerging pathogens' has been applied, particularly to the genera *Yersinia*, *Aeromonas* and *Plesiomonas*. Of interest, as with *Listeria* spp., is the ability of some of these bacteria to grow at refrigeration temperatures. This means, of course, that in non-sterile foods in which inhibition of microbial growth is not achieved chemically, refrigeration as the principal means of controlling food-poisoning bacteria may not provide adequate protection if these organisms are present.

—— 2.3 ——
ANTIGENIC STRUCTURES

The key concept in immunological detection of food-poisoning bacteria is, as with other analytes, the specific binding of antibody to antigen. The latter term is widely used although, more correctly, antibody binding is to one or more epitopes (antibody-binding sites) which are usually part of a larger immunogenic structure. The term 'antigen' will, however, continue to be used frequently in this book, because bacterial surface 'antigens' will usually contain multiple epitopes. In the case of monoclonal antibodies, binding is to a single type of epitope. Epitopes are thought to be approximately 5–15 amino acid residues in size. The structure used to raise antisera is important to the specificity of the final assay, as will be seen later, and on a large entity such as a bacterial cell many possible structures are to be found; this is also true, although to a lesser extent, of enterotoxins.

Antibodies are produced *in vivo* in response to compounds perceived as foreign, of molecular weight generally more than 1000; large peptides or proteins can be extremely immunogenic but it is possible to raise antibodies to polysaccharides and other large molecules. Small molecules with molecular weights of less than 1000 need to be covalently coupled to a large molecule (usually a protein – see Appendix 2) to produce an immune response. Many of the different antibodies so produced will be to the carrier protein molecule but a proportion will recognize the coupled small molecule (known as a hapten) alone. In this way specific antibodies to molecules with a molecular weight as small as 100 have been raised.

2.3.1 Bacterial toxins

Bacterial exotoxins and endotoxins consist mostly of high-molecular-weight materials, principally polypeptides, lipids and polysaccharides. As such, they are generally highly immunogenic, readily stimulating antibody production. The problem, of course, is toxicity towards the animal immunized. There are several possible ways around this. The material can be detoxified before injection (at the risk of altering the structure of potential epitopes), or a non-susceptible animal species can be used to raise antisera. The latter approach presents obvious problems if the animal has to be one not normally used for antibody production. A further possible method is to follow an immunization schedule that starts with very small amounts of toxin, and then gradually increases the dose. In this way the animal can develop a tolerance to the toxin (Lee and Morgan, 1990). Detailed methodology for the conversion of toxins to non-toxic material (toxoid) is beyond the scope of this book; briefly, treatment with formaldehyde or heat is commonly used (Lee and Morgan, 1990). Although there are similarities between some toxins, e.g. *E. coli* LT and cholera toxin, it is possible to raise specific antisera. Commercially available assay kits for several toxins will be discussed later in the book.

2.3.2 Bacterial cells

A bacterial cell contains numerous potential antigenic structures, many of which will not be specific to that species. Indeed, classically these have been used to show relatedness of species within a higher taxon, and antigens are commonly shared between closely related species. Conversely, species can be subdivided into serotypes based on the spectrum of antigens possessed.

From the standpoint of developing an immunoassay for a bacterial cell, antigens exposed on the cell surface or excreted into the surrounding environment (such as exotoxins) present the best target. if internal antigens are used, then a means of gaining access to these has to be included in the assay protocol, adding an extra step. Additionally, if chemical or heat treatments are to be used for extraction of antigens, then there is also the risk of unpredictable changes to the target epitopes.

If whole bacterial cells are used as immunogens to raise antisera in animals, then the resultant polyclonal antisera will contain antibodies to many surface structures. Before the advent of monoclonal antibody technology, unwanted cross-reactivity was removed from serum, at least in part, by the process of absorption, in which the serum was exposed to cells of the 'unwanted' bacterial species; this allowed binding of 'unwanted' antibody to the cells, consequently freeing the serum of this reactivity. This was not, however, a very satisfactory or convenient process, but could produce sera that were capable of confirming the identity of a purified culture, or for classifying species into serotypes. Use of these sera on mixed cultures is not possible and they do not make good reagents for immunoassays.

Monoclonal antibody technology, however, allows the production of hybridoma clones secreting antibody of a single specificity; screening of hybridoma cell lines against the organism of interest, and against other organisms for cross-reactivity, enables a heterogeneous starting material, such as whole bacteria, to be used as an immunogen. Nevertheless, if a particular bacterial cell surface component known to be characteristic of the species is used as the immunogen then the chances of early selection of a specific antibody-producing hybridoma are increased. In fact, useful and specific polyclonal antisera can be produced by immunization with a cell surface component, if a suitable specific structure is known and can be purified free of all contaminating material, including components of culture media. Production of polyclonal and monoclonal antibodies is described later in the book.

The principal surface structure on Gram-negative cells (Figure 2.1), in particular of the family Enterobacteriaceae, is the lipopolysaccharide (LPS) known historically as the 'O' or somatic ('body') antigen. Indeed, this material is so abundant that it acts as a toxin and is believed to be responsible for many of the symptoms of infections with Gram-negative organisms. For present purposes, the important constituents of this material are the glycolipid (lipid A) and polysaccharide moieties, which

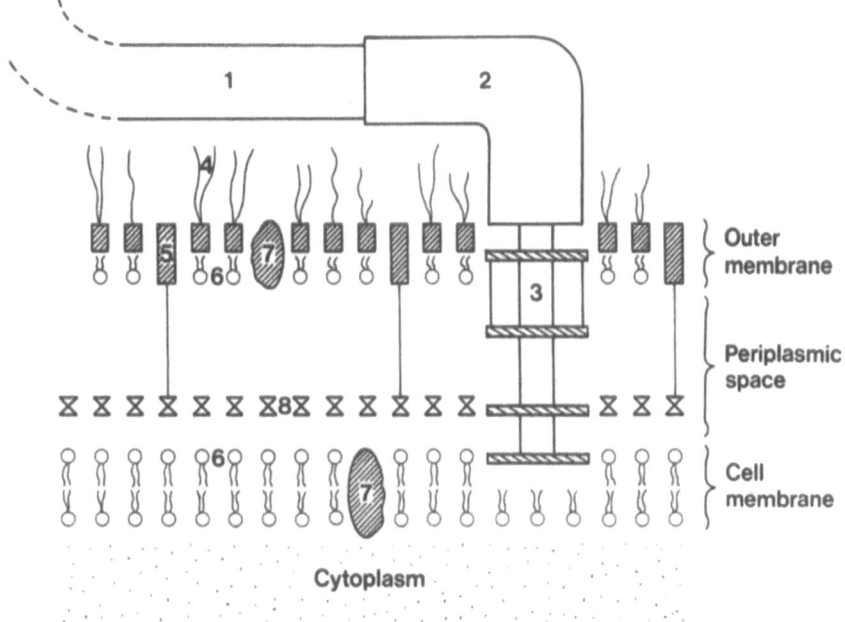

Figure 2.1
Diagrammatic representation of the structure of a generalized Gram-negative cell envelope. (1 = flagellar filament 'H' antigen; 2 = flagellar hook; 3 = flagellar basal body; 4 = lipopolysaccharide 'O' antigen chains; 5 = lipoprotein; 6 = phospholipid; 7 = protein; 8 = peptidoglycan)

are the outermost components of the complex. Lipid A alone in solution has little activity but, when extracted together with the polysaccharide chain, it is a potent toxin. The polysaccharide chain has a core section which appears to be common to closely related species, and an O-specific chain with a highly variable structure of hexoses and deoxyhexoses in repeating oligosaccharide units which are responsible for the specificity of the chain. The immunodominant sugars of some groups of O specificity are of unusual structure, e.g. paratose, abequose and tyvelose in salmonella groups A, B and C, and D respectively. 'Rough' mutants of several genera exist, named from their appearance on agar plates, which have various levels of incomplete polysaccharide chain.

The extraction method employed depends on which layers of the LPS complex are required. An extremely immunogenic preparation, which contains LPS plus protein and phospholid components from the cell envelope, is extracted with trichloroacetic acid (see Appendix 3). Hot phenol–water extraction gives LPS with slight outer membrane protein contamination whereas ethylenediamine tetraacetic acid (EDTA) treatment gives pure LPS (see Appendix 3). Some of these methods may co-extract the so-called enterobacterial common antigen from members of the Enterobacteriaceae family. Constituent sugars can be released from the polysaccharide, if required, by standard techniques of carbohydrate chemistry and, because these are haptenic structures (see Section 2.3), they can be used to form conjugates with proteins (see Appendix 2) to give an immunogenic complex.

A rather cruder method of 'extraction' is to rely on the heat stability of LPS on the cells. When a saline suspension of whole cells harvested from a culture is boiled for 2.5 h, cell-surface protein antigens are denatured leaving a preparation that is largely LPS. This is the method that has long been used to produce sera for classic serotyping schemes; cross-reactivity was reduced by absorption as described previously.

Whereas the LPS layer extends from the cell envelope to a distance of the order of 10 nm, the flagellar filaments of flagellated species can be as long as 20 μm from the surface. This length, together with the protein nature of the filaments, ensures a strong immune response to flagella, which historically have been termed 'H' antigens. (As an aside, the designations H and O antigens were derived from the German terms *Hauch* and *Ohne Hauch*, describing the appearance of spreading and non-spreading colony forms.)

The structure of flagella has been studied the most in *Salmonella typhimurium*. A flagellum consists of three parts: a basal body that is incorporated in the cell envelope, a hook that emerges a short distance, and a filament that forms the most lengthy and visible part of the structure. The filament is a polymer composed of repeating protein subunits with molecular weights of the order of 40 000–60 000 arranged, in some species at least, in a helical shape with a hollow core. Dissociated monomers have the ability to re-assemble into filaments visually indistinguishable from native flagellin, although differences in antibody binding have been reported in salmonellae between native and re-polymerized flagellin (Ibrahim *et al.*, 1985). Flagella can be removed from whole cells by mechanical shearing which also removes a proportion of hooks and basal bodies together with some cell fragments; purification is then carried out by differential centrifugation (see Appendix 3). Flagellin protein subunits can be prepared by conventional protein chemistry. It should be noted, however, that epitopes appear to be unmasked in monomeric flagellin that are not available for binding in native flagella (Ibrahim *et al.*, 1985).

A better method for preparation of pure flagellin protein free of cell fragments is brief solubilization of pH 2.0 with removal of insoluble material and ammonium sulphate precipitation (see Appendix 3).

Before the development of monoclonal antibodies, it was suggested (Langman *et al.*, 1972) that for salmonellae antigenic determinants existed on flagella that were outside the historical (Kauffman–White) H-antigen typing scheme and this has been confirmed in the authors' own work – there are monoclonal antibodies (i.e. those that bind only to a single epitope) to flagellin protein, the cross-reactivities of which do not correspond to that scheme.

Some species, particularly *Salmonella* spp., exhibit two alternative forms or phases of H-antigen expression. Each individual cell has H specificity of one phase only but, in a completely single-phase population, cells of the alternate phase arise with a frequency of about 10^{-4}. Cells of a single phase can be isolated from a biphasic population by immobilizing the alternative phase with specific antiserum in a soft agar, which allows cells of the opposite phase to spread across the surface of the agar.

As with traditional 'O' antisera, crude sera to H antigens can be raised by immunization with whole cells, using 'rough' mutants lacking the LPS antigens. Again, absorption of unwanted cross-reacting antibodies is necessary.

Certain proteins of the Gram-negative outer membrane may be exposed on the cell surface (OMP) and may be genus, species or serotype specific. OMP can be prepared by detergent extraction (see Appendix 3) and purified by standard protein chemistry techniques.

One further type of surface antigen is exposed on Gram-negative cells – the pili (or fimbriae). These are protein filaments, but they are shorter, thinner and more rigid than flagella, and are involved in adhesion of the cells – possibly part of the pathogenic mechanism in some species.

The glycopeptide layer of the cell wall does not appear to be exposed in Gram-negative cells because these organisms are not susceptible to formation of spheroplasts (forms that have lost cell wall rigidity) by lysozyme treatment, unless the LPS layer is deficient. However, in Gram-positive organisms the cell wall is thicker and

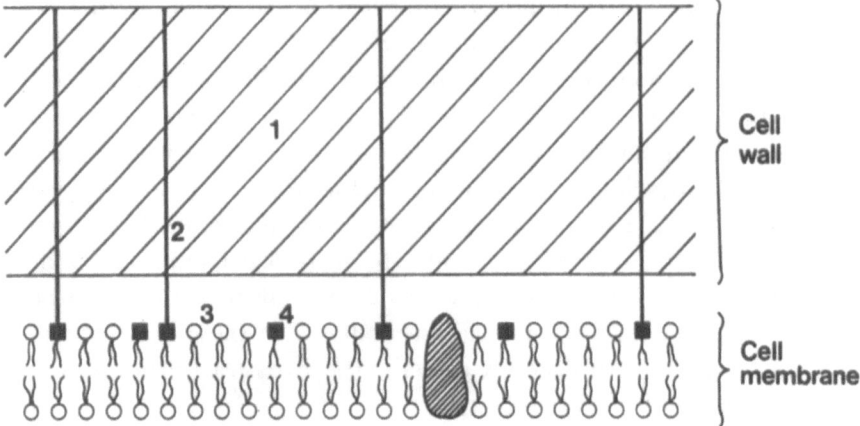

Figure 2.2
Diagrammatic representation of the structure of a generalized Gram-positive cell envelope. (1 =
amorphous cell wall containing teichoic acids and peptidoglycan; 2 = lipoteichoic acid; 3 =
phospholipid; 4 = glycolipid)

external and thus a potential target for antibody binding (Figure 2.2). The surface components of the wall are a group of polymeric alcohols known as teichoic acids; these are important in the immunogenicity of Gram-positive cells and may be species specific. Extraction of teichoic acids is by mechanical disruption of the cell walls and solubilization in trichloroacetic acid (see Appendix 3).

The remaining external cellular material to be considered is the slime or capsular polysaccharides loosely bound to the cells of several species. This material is, however, variable even within a species, it can be lost without apparent effect on the organism and it may be expressed only under certain environmental conditions. These layers, therefore, do not appear to make a good target for rapid antibody-based assays.

As discussed in earlier sections, several food-poisoning species are spore formers. The surface layers of spores are rather different in composition and structure to those of vegetative cells and antibodies against cells would not necessarily cross-react with spores. However, in many situations where spores are present, then vegetative cells are also likely to be present, so this may not be a problem in practice.

As mentioned earlier, an ideal target for the antibody in an assay would be an immunogenic compound excreted by the cell into the surrounding environment. The particular case of known exotoxins has already been discussed. However, there is still the possibility that other specific compounds, e.g. enzymes or lysins, may be excreted by the organism of interest. This is an area that has been little explored.

Before concluding this section, bacteriophage receptors as potential specific epitopes on bacterial cells are worth a brief mention. It is known that there is a multiplicity of phages of both very broad and very narrow (often subspecies or subserotype) specificities, which have receptors (i.e. specific attachment sites), in almost all bacterial cell surface structures. The sites are small and probably depend on both conformation and composition of the structure; however, these criteria also apply to antibody-binding sites and thus antibodies to phage receptors are entirely feasible. Indeed, it is possible that existing specific monoclonal antibodies recognize these sites in some cases; this idea could be tested in studies of inhibition of phage susceptibility.

BIBLIOGRAPHY

Bergey's Manual of Systematic Bacteriology (1984) Ed. J. G. Holt, Williams and Wilkins, Baltimore, USA

Topley and Wilson's Principles of Bacteriology, Virology and Immunity (1990) Eds M. T. Parker and L. H. Collier, Edward Arnold, London, UK

General Microbiology (1987) Eds R. Y. Stanier, J. L. Ingraham, M. L. Whellis and P. R. Painter, Macmillan Education Ltd, Basingstoke, UK

Microbiology (1990) Eds B. D. Davis, R. Dulbecco, H. N. Eisen and H. S. Ginsberg, J. B. Lippincott Company, Philadelphia, USA

Food-Borne Infections and Intoxications (1979) Eds H. Riemann and F. L. Bryan, Academic Press, New York, USA

Microbial Cell Walls and Membranes (1980) Eds H. J. Rogers, H. R. Perkins and J. B. Ward, Chapman and Hall, London, UK

Methods in Immunology and Immunochemistry Volume 1. Preparation of Antigens and Antibodies (1967) Eds C. A. Williams and M. W. Chase, Academic Press, New York, USA

CHAPTER

—3—

Historical Aspects

— 3.1 —
USE OF ANTIBODIES IN MICROBIOLOGY

A central problem in the use of antibodies as a test reagent is the recognition that the antibody–epitope reaction has occurred; much more will be said about this later in the book. All early work with antibodies and bacteria was based on direct observation of the visible lattice or network formed from the cross-linking of bacterial cells by the bivalent antibodies previously invisible to the naked eye. The *in vitro* phenomenon of these agglutination and precipitation processes was known at the beginning of the twentieth century. It was realized early on that immune animal sera consisted of mixtures of binding capacities (i.e. polyclonal) and the *in vitro* use was largely confined to 'typing' (i.e. taxonomic allocation of) pure cultures of bacteria. The precipitation techniques of immunodiffusion and immunoelectrophoresis for soluble antigens in a semi-solid matrix became quite sophisticated and were used for detailed studies of the occurrence of particular antigens in bacteria. Agglutination reactions, i.e. those that occur when an antiserum cross-links to antigens fixed to the surface of a cell, are still widely used in the form of slide tests for serotyping bacteria; a degree of sophistication has been introduced with the use of latex particles, sometimes colour coded, as antibody carriers (see Section 5.8.2).

A big advance was made in 1941 with the introduction of labelling of antibodies with fluorochromes – compounds that fluoresce when stimulated by particular wavelengths of light. By washing away unbound fluorescent antibody from a preparation of bacteria, binding of antibody to cells could be seen microscopically when illuminated with light of the correct wavelength.

All these methods use antibody at far higher concentrations and are very much less sensitive than the ELISAs which will be described later.

—— 3.2 ——
TRADITIONAL CULTURE METHODS FOR BACTERIA

When proposing a new technique of analysis, comparisons have to be made with an accepted benchmark in order to assess the performance of that technique. In the case of detection of food-poisoning bacteria, the comparison is with the existing cultural methods. Often, as will be seen, this is by no means a perfect standard.

Culture and identification of bacteria from an environment containing a single type of organism, even if present in small numbers, are generally straightforward and relatively fast. However, ecological niches supporting only a very limited range of bacteria are usually of a homogeneous nature, and most foods do not fall into this category. The diverse composition and range of structures within food, together with the varied origins of the food components, leads to the presence and potential for growth of a wide range of bacterial types. Among this population, food-poisoning organisms may form only a small proportion in the starting material. An additional problem is that any non-sterilizing processing of the food materials involving heating, drying or freezing may lead to sublethal damage to those members of the population that are not completely killed by the process. These organisms may have membrane damage or breaks in the nucleic acid polymers which are repairable in favourable conditions to give a fully competent cell. However, in the injured state bacteria are sensitive to environmental stress and isolation techniques must allow for this if their presence is not to be overlooked.

A technique for detecting small numbers of a given bacterial species in a mixed population, known as selective enrichment culture, is based on the principle of providing conditions more favourable (or less unfavourable) for growth of the target organism than for the competing flora. The agent of selectivity can be physical, such as pH, temperature, osmotic pressure or a particular composition of atmospheric gases, or it can be a chemical compound added to the culture medium; these compounds are usually dyes, antibiotics or other inhibitory chemicals.

The toxicity of these chemical selective agents presents at least three problems. First, the sensitivity of sublethally damaged cells, as discussed above, necessitates a non-selective cultural stage to allow resuscitation to fully viable cells prior to exposure to the selective agent; this lengthens the procedure. Secondly, within a genus such as *Salmonella*, not all strains respond equally favourably to the selective conditions and thus more than one selective culture, based on different principles, has to be used to ensure complete coverage. Thirdly, none of the agents is totally selective, i.e. they only partially suppress the competing flora, allowing the target cell to grow, but at a reduced rate of growth. Thus, a further cultural stage is necessary – the diagnostic agar plate. Even then confirmatory tests on suspect colonies are often required.

—— 3.3 ——
DETECTION OF TOXINS BY TRADITIONAL METHODOLOGY

Traditional methods for toxin assay are divided into two types: *in vivo* and *in vitro*. The former involve animal challenge tests and, depending on the toxin type, use various skin or intestinal injection techniques. *In vitro* testing for toxins was an area involving early use of antibodies (antitoxins) by methods mentioned in Section 3.1; various immunodiffusion, immunoelectrophoretic and haemagglutination techniques were developed. Tissue culture using toxin-sensitive cells has also been developed as a replacement for animal tests. All of these methods are laborious and difficult to apply directly to complex foodstuffs.

—— 3.4 ——
A NEW PERSPECTIVE ON DETECTION OF BACTERIA?

It was shown previously that the problems thrown up by the use of inhibitory chemicals as selective agents in culture are numerous, and alternative means of detecting a target organism could result in considerable savings in time, labour and materials. The binding of antibody to epitope can be very specific and, with a suitable antibody, this offers far better selectivity in a mixed population than would ever be achieved by chemical or physical means, without the parallel problems outlined earlier. In the past, detection of single cells has relied on the amplification brought about by multiplication of the single cell to the point at which visible colonies form on a suitable diagnostic agar; this amplification is of the order of 10^7- to 10^8-fold. In theory, binding of antibody to a single bacterial cell can be manipulated to give sufficient signal to recognize that event without the time lag necessary for multiplication of the organism; an additional and very considerable advantage is that the binding event will only take place with a known target according to the specificity of the antibody used. This dual ability of antibodies to provide both specificity and sensitivity has offered a new direction in detection of bacteria, away from methodology first developed 100 years ago, and a more practicable method of toxin assay.

BIBLIOGRAPHY

Microbiology (1991) Eds B. D. Davis, R. Dulbecco, H. N. Eisen and H. S. Ginsberg, J. B. Lippincott Company, Philadelphia, USA

Topley and Wilson's Principles of Bacteriology, Virology and Immunity (1990) Eds M. T. Parker and L. H. Collier, Edward Arnold, London, UK

General Microbiology (1987) Eds R. Y. Stanier, J. L. Ingraham, M. L. Whellis and P. R. Painter, Macmillan Education Ltd, Basingstoke, UK

Bacteriological Analytical Manual (1978) Association of Official Analytical Chemists, Washington DC, USA

Microorganisms in Foods (1978) Eds R. P. Elliot, D. S. Clark, K. H. Lewis, H. Lundbeck, J. C. Olsen and B. Simonsen, ICMSF, University of Toronto Press, Canada

Compendium of Methods for the Microbiological Examination of Foods (1976) Ed. M. L. Speck, American Public Health Association, Washington DC, USA

Handling Laboratory Microorganisms (1991) C. Penn, Open University Press, Buckingham, UK

CHAPTER
—4—

Food as an Assay Matrix

In the sense used here, the matrix is the complex of food components that is to be analysed for the presence of the target analyte. In any assay procedure, not just those based on antibody binding, the food matrix can interfere and produce an erroneous result. When assaying for the presence or absence of a target, food has the potential to give 'false-positive' results, i.e. those that are positive when the true result is negative, and also the converse 'false-negative' results. This, of course, begs the question: What is the 'true' result? As mentioned previously, this can only be assessed in relation to a standard, well-tried method, which may in itself be less than perfect.

—— 4.1 ——
PHYSICAL AND CHEMICAL PROPERTIES OF FOOD

Physically, most foods are solids or semi-solids. Reference was made earlier to the theoretical detection of the binding of antibody to a single cell; the physical state of many foods may, of course, make finding a particular single cell in the matrix extremely difficult. Nevertheless, the analytical standard for many food-poisoning bacteria is that the method should detect one cell in a 25 g food sample; in reality, this translates to being sure of the *absence* of any cells in that sample. A variety of blending and extraction methods exists which enables a food sample to be reduced to a slurry with a solvent, incorporating an emulsifying agent if necessary, making handling easier and access to the target cell more likely.

The only other physical property of foods likely to be of importance is pH. Antibody binding takes place over a wide pH range ($\sim$ 4–9 for many polyclonal preparations) and this is unlikely to be affected by the kind of pH levels common in foods; in any case, assays are generally optimized by being carried out in buffer.

The principal chemical components of foods are carbohydrates, fats, protein and water; in the present context, minerals and vitamins are of less importance. Water alone would be the best possible assay matrix, so the components that are most likely to interfere with an immunoassay are the first three mentioned above. The levels of these components in foods which might be vehicles for microbiological poisoning vary widely; as examples, whole eggs are approximately 75% (w/w) water, 12% protein and 11% fat with only a trace of carbohydrate, whereas chocolate is approximately 2% water, 8% protein, 30% fat and 60% carbohydrate. In the experience of the authors' group, milk can be a particularly troublesome assay matrix.

Several classes of interference are possible. First, large molecules may simply cause steric hindrance of the antibody–epitope binding event, interfering with the kinetics of the reaction, possibly reducing the assay signal and adversely affecting sensitivity. Secondly, non-specific reactions may occur due to non-immunological binding of the antibody to food proteins or to protein A produced by *Staph. aureus* (the latter effect might be less of a disadvantage in the present context). Thirdly, the natural presence in foods of either enzyme capable of catalysing the assay substrate, or inhibitors of that enzyme, can lead to erroneous final colour development in assays based on that principle if the intermediate washing stages are not completely effective. Matrix interference will be covered in more depth in Chapter 7.

When immunoassays are used for soluble analytes, the above problems can often be circumvented by a solubilizing extraction procedure prior to analysis, but this is not applicable to bacteria, unless surface antigens could be so extracted. However, cell separation and concentration systems are being developed and are discussed later.

—— 4.2 ——
MICROBIAL FLORA OF FOOD

As mentioned in Section 3.2 it is usual to find a number of types of bacteria in a food rather than a single species; the exception to this might be in a processed food where, due to a faulty process, a very small number of a particular resistant type, e.g. a spore-bearing organism, survive in an otherwise sterile food. These might then germinate and outgrow to give a pure culture in the food.

Assuming that the food under examination contains other organisms, in addition to the target food-poisoning species, then the presence of these can influence immunoassays in at least two ways. First, if the assay procedure includes a cultural stage then the food flora may compete with the target organism for nutrients, lower the pH of the medium sufficiently to prevent growth of the target cell, or produce toxic metabolites such as antibiotics and bacteriocins. Secondly, closely related organisms in the natural food flora may share antigenic structures with the food-poisoning species; this is particularly true with members of the family Enterobacteriaceae to which several food-poisoning genera belong. It is essential that the antibodies being used are specific for the target organism.

There seems to be little point here in listing all the possible components of the microbial flora of foods. Suffice it to say that the organisms are often grouped by metabolic activity or characteristics rather than on a strictly taxonomic basic. For instance, the groups commonly considered include: lactic acid-forming bacteria, proteolytics, lipolytics, thermophiles and psychrophiles; some judgements can be made of the possible effect of members of the flora on an assay system from these characteristics.

BIBLIOGRAPHY

Food Science – A Chemical Approach (1970) B. A. Fox and A. G. Cameron, University of London Press, London, UK

Principles of Food Science – Part I, Food Chemistry (1976) Ed. O. R. Fennema, Marcel Dekker Inc., New York, USA

Microbial Ecology of Foods – Volume 2, Food Commodities (1980) Eds J. H. Silliker, R. P. Elliott, A. C. Baird-Parker, F. L. Bryan, J. H. B. Christian, D. S. Clark, J. C. Olsen and T. A. Roberts, ICMSF, Academic Press, New York, USA

CHAPTER
—5—

Antibody Technology

In virtually all forms of immunoassay, three principal requirements have to be met. First, one or more antibodies of the desired specificity are needed, and targets for this specificity have been discussed. Secondly, it is necessary during the assay to separate out those antibody and/or analyte molecules that have not bound to targets from those where specific binding has taken place. Finally, the occurrence of this binding event must be recognized and quantified; to do this a macroscopic signal has to be generated by the event and measured in some way. These three requirements are the subject of this chapter.

—— 5.1 ——
IMMUNIZATION PROTOCOLS

It should be stated clearly here that the use of animals to raise antibodies must be regarded as a privilege and not a right. That said, the procedures involved in antisera production cause very little distress to the animals used, being a natural *in vivo* process.

In the UK, experimental use of animals is covered by the Animals (Scientific Procedures) Act 1986 and both project licences, giving general permission for the work, and personal licences, for each individual worker, are needed. Premises and animals are subjected to inspection by the Home Office and an appointed veterinary surgeon, under whose care the animals remain. The Home Office needs to be satisfied that each worker is fully competent to carry out the required procedures.

Although many of the structures likely to be used as immunogens for production of antibodies to bacteria would be immunogenic in themselves, it is usual to make a preparation in an adjuvant, which increases response to an immunogen by generally stimulating the immune system. The most commonly used adjuvants are Freund's water-in-oil emulsions; the immunogen is dissolved in the water phase and diffuses slowly into the tissues, prolonging the stimulus. An additional effect of these emulsions is to retard destruction of the immunogen. Freund's 'complete' adjuvant also contains dead mycobacteria which provoke an inflammatory response, enhancing antibody formation. However, due to the risk of lesions at the injection site, complete Freund's is used only for the first injection and incomplete, i.e. without mycobacteria, is used for subsequent injections. Alternative adjuvants are available and a saponin-based preparation, Quil-A (Superfos, Denmark), has been used with success for bacterial cell extracts. Care must be taken when handling adjuvants, particularly Freund's, because hypersensitive reactions can occur if the adjuvant gains entry through the skin or by breathing aerosols created during manipulation of the material.

The response of the animal to the immunogen will vary with the dose, the nature of the immunogen and adjuvant, the route of administration and even the individual. However, it can be said in general that the primary response, i.e. that to the first injection, is smaller and more delayed than that to subsequent injections under the same conditions. Mainly rabbits have been used for production of polyclonal antisera but chickens have been used, which have the advantage that antibody is excreted in the eggs. For rabbits, a general procedure is to give a first dose of 100 μg protein in 1 ml Freund's complete adjuvant (Difco – a 7:3, v/v mix of adjuvant and sterile isotonic saline, vigorously mixed to form the emulsion). This is divided between intramuscular injections of the two hind thighs (total 0.5 ml) and two subcutaneous injections on the back (total 0.5 ml). A second dose, in Freund's incomplete adjuvant, is given after 6 weeks. Small areas of skin are exposed by shaving at the intended injection sites, and the areas swabbed thoroughly with an alcohol-based cleaning agent prior to injection. Blood (10 ml in a heparinized tube) is taken from the marginal ear vein twice during a period 8–14 days after the second injection; usually bleeding at 9 and 13 days will cover the peak of the antibody response. Blood cells are removed by centrifugation and the serum stored at −20°C. Subsequent injections and bleeding cycles can be performed at 6-week intervals.

For production of monoclonal antibodies, female BALB/C mice are immunized with 50 μg protein in 0.2 ml adjuvant, injected intraperitoneally on a schedule similar to that for rabbits. A single test bleed is taken from the tail and, if the animal has responded (see Appendix 7), spleen cells are used for lymphocyte fusion (see Section 5.3) 4 days after a subsequent injection. Very recently, the UK Home Office has issued a recommendation that only subcutaneous injection sites be used in protocols for antibody production.

—— 5.2 ——
POLYCLONAL ANTIBODIES

Antibodies are serum proteins classed as immunoglobulins and are produced from B lymphocytes in response to the presence of 'foreign' material, i.e. material not recognized as 'self'. These cells have receptor sites on their surface which are unique to each clone. When stimulated by the binding of an immunogen, the particular cell line proliferates and produces antibody of the same specificity as the receptor. A number of genetic mechanisms allow an individual to produce at least 10^6 different antibodies and a large immunogenic structure, with many different epitopes, will stimulate many different lymphocyte clones, producing a mixed, or 'polyclonal', antibody response. Serum from animals immunized with such structures will be polyclonal.

Chemically, antibodies are large protein molecules called immunoglobins (Ig). The predominant type is IgG which has a molecular weight of about 160 000 and a structure resembling a 'Y' shape (Figure 5.1), formed from two heavy chains and two light chains. The chains have variable regions at the tips of the two arms (the Fab region) which are responsible for the specificity of the binding site. The two arms forming the Fab region comprise about two-thirds of the weight of the molecule; the remaining one-third, the 'tail', is the Fc region which is largely constant for all antibodies of one class and one animal species. IgG is divided into four subclasses, and there are four other classes of antibodies, IgA, -D, -E and -M. Depending on the class, these exist as monomers, dimers or pentamers. IgM is usually the first antibody to

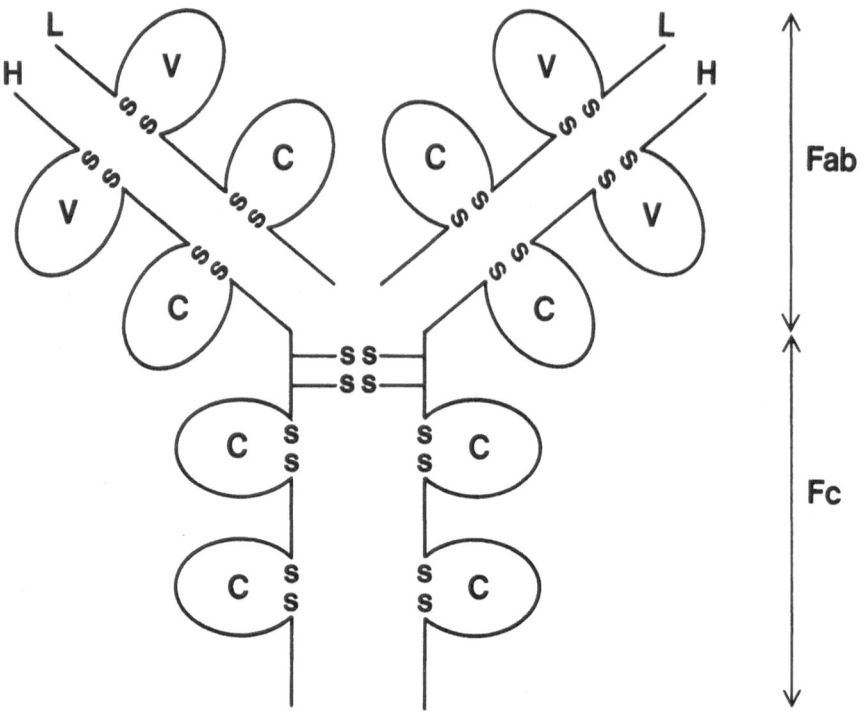

Figure 5.1
Diagrammatic representation of immunoglobulin G molecule. (SS = disulphide bridges; H =
heavy chain; L = light chain; V = variable regions; C = constant regions; Fab = antigen-
binding fragment; Fc = crystallizable fragment)

appear in the serum, where it exists as a pentamer with a total of 10
binding sites, making for efficient binding and subsequent lysis of
invading organisms, or agglutination of particulate targets. IgG is the
most abundant serum immunoglobulin, existing as a monomer, and
appears rather later than IgM, although it continues to increase in
concentration after IgM levels have stabilized or started to fall.
Prolonged stimulation can produce detectable serum concentrations of
IgA, the main immunoglobulin of other body fluids.

Antibodies are produced in response to the presence of a foreign
protein. This protein could itself be an antibody, which of course
possesses many epitopes, giving rise to various types of anti-antibodies
classified as follows. An anti-isotype is an antibody to the class of
immunoglobulin, i.e. the general structure of the light and heavy

chains. An anti-allotype is an antibody to the site characteristic of the individual that produced the first antibody, whilst an anti-idiotype binds to the region that is chemically unique to that antibody. Some anti-idiotypes will actually be antibodies to the binding site of the first antibody, and would thus compete with the specific target for those sites. Such anti-idiotypes have been used experimentally as vaccines and are also in the early stages of development for use as reagents in immunoassays where the target has a specific function, such as the active site of a toxic molecule. The anti-species antibodies discussed later are also a group of anti-antibodies.

Methods for assessing the concentration, specificity and class of antibody present in a polyclonal serum (or monoclonal culture supernatant) will be discussed later in this chapter and in the appendices.

—— 5.3 ——
MONOCLONAL ANTIBODIES

The principal advantage of the use of monoclonal antibodies is that of defined specificity; unlike the mixed antibody preparations discussed above, antibody that binds to only a single epitope can be produced in virtually unlimited amounts in cell culture. An additional considerable advantage is that, apart from the initial immunized mouse, no further animal work is required.

Unfortunately, B lymphocytes will not multiply in artificial culture and therefore production of antibody *in vitro* was not possible until the discovery by Kohler and Milstein (1975) of a method of fusing these cells to a cell line that would replicate; the hybrid cells (hybridomas) so produced retain the property of antibody production from the lymphocyte 'parent' and yet they are immortal, i.e. they will divide endlessly in culture given the correct conditions. Myeloma cells are used as the immortal cell line and many well-characterized lines which have been adapted to grow in artificial culture are available. They are derived originally from a cancerous condition of antibody-secreting cells, usually in BALB/C mice and are thus compatible with lymphocytes taken from the spleen of immunized mice of the same strain. The basis of the fusion process is that multi-nucleated cells are formed under the influence of an agent that allows the cell membranes to fuse; subsequent cell division produces daughter cells sharing genetic

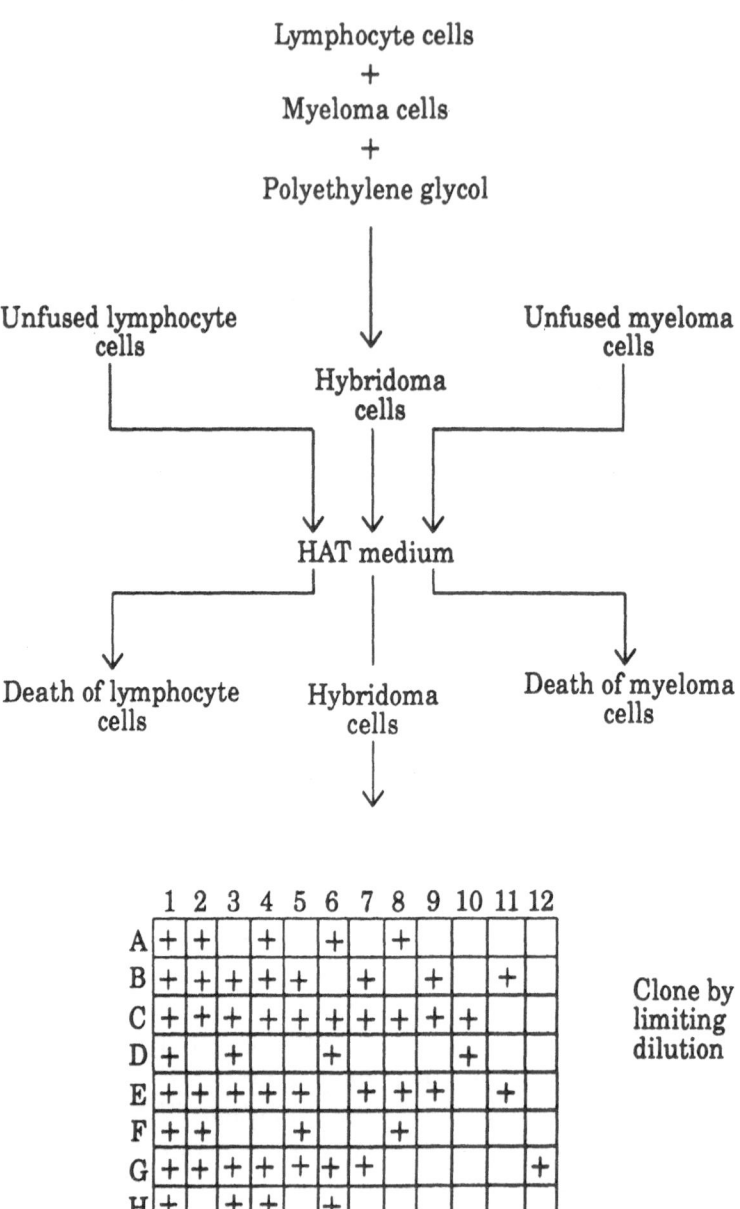

Figure 5.2
An outline of monoclonal antibody production. (HAT = hypoxanthine, aminopterin, thymidine; + = growth of hybridoma in culture plate)

material from the parent cells. The original fusing agent used by Kohler and Milstein was the Sendai virus, but this has been superseded by polyethylene glycol.

Several fusion combinations are possible in a mixed suspension of spleen cells and it is necessary to select out the desired hybrids (Figure 5.2). Unfused spleen cells will not survive for more than a few days in culture and death of unfused myelomas can be brought about by use of a selective cell culture medium containing hypoxanthine, aminopterin and thymidine (HAT medium). Aminopterin blocks the main biosynthetic pathway leading to DNA synthesis, but cells containing hypoxanthine–guanine phosphoribosyltransferase (HGPRT) can utilize hypoxanthine and thymidine in an alternative pathway. If the myeloma cell line is prepared such that it lacks the enzyme, then only hybridoma cells, which have gained the enzyme via fusion of a myeloma cell with a spleen cell, can grow. It is essential that stocks of myeloma cells are checked at intervals for sensitivity to HAT medium; this can be done by exposure to 8–azoguanine $(20 \ \mu g \ ml^{-1})$ during growth which eliminates any cells possessing HGPRT.

Having eliminated all unwanted cell types, lines secreting useful antibody are selected by screening culture supernatant against the desired analyte using an ELISA, as discussed later in this chapter. The cultures are then cloned, i.e. single cell lines are selected, by repeated limiting dilution and further culture until it is statistically probable that the cell lines are monoclonal.

Cell culture has to take place under sterile conditions and all work should be carried out in laminar-flow sterile air cabinets. Radiation-sterilized tissue culture-grade disposable flasks and other culture vessels are available from several suppliers. Streptomycin and penicillin are routinely incorporated in culture media as an extra protection against microbial contamination. However, on occasions cell cultures do get contaminated and a great deal of valuable work can then be lost. The routine of cell culture can be tedious and time-consuming, but can be semi-automated by the use of robotic equipment such as the 'Biomek' (Beckman Instruments – Figure 5.3).

The need to preserve samples of cells at all stages of production of a monoclonal line cannot be over-emphasized. In addition to the possible intervention of microbial contamination, cell death can occur for other reasons and, in any case, long-term preservation of the final cell line is necessary. Cells stored in vials in liquid nitrogen can remain viable for several years.

Figure 5.3
A computer-controlled robotic system for handling hybridoma cell culture, installed in a sterile-air cabinet

Antibody is excreted into the culture medium and the titre can also be determined by ELISA. The alternative method of antibody production is not used – this involves injection of the hybrid cells into mice or rats to induce tumour formation; ascites fluid associated with tumour growth can contain antibody levels of 5–50 mg ml^{-1}, much higher than that produced in culture (5–50 μg ml^{-1}). However, for several reasons it is felt to be preferable to accept this lower level of *in vitro* production, enlarging the scale as necessary to produce the required quantity.

Most of the antibodies in a polyclonal serum will be IgGs of average affinity; chance thus dictates that most monoclonal lines selected will have less than the highest affinity. Because monoclonal antibodies are chemically identical, their binding properties may be affected by small changes in conditions to a much greater extent than is apparent with a polyclonal serum; in fact, the large number of different antibodies in a polyclonal preparation will probably give binding in a wide range of conditions, with binding transferring from one antibody to another as the conditions change.

Detailed protocols for monoclonal antibody production are given in Appendix 4.

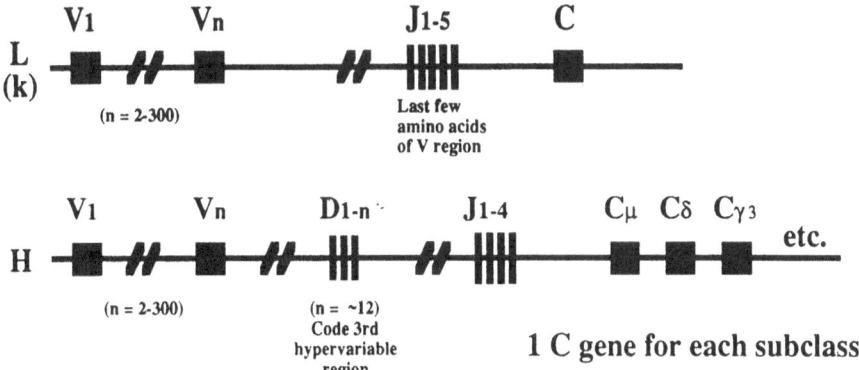

Figure 5.4
Organization of antibody gene clusters for the heavy (H) chain and light kappa (L, k) chain on different chromosomes (V, J, C, D = antibody gene clusters)

—— 5.4 ——
ANTIBODY PRODUCTION BY GENETIC MANIPULATION TECHNIQUES

Recently, molecular biology techniques have enabled the production of antibodies and antibody fragments of widely varying structures from a variety of different expression hosts. To understand these procedures, it is necessary to look at the genetics of antibody production. How does the mammalian immune system produce an estimated 10^8 or more different antibodies to cope with the enormous variety of antigens that it may encounter? An antibody molecule is made up of several domains, one variable (V) and three constant (C) domains for each heavy chain and one variable and one constant domain for each light chain. A large part of the variable domain forms the framework region, which is actually fairly constant and which acts as a backbone for three hypervariable loops (complementarity determining regions or CDRs). It is these CDRs, three from the heavy chain and three from the light chain, that are spatially arranged to form the binding site. The domain structure of the antibody is mirrored at the level of the gene with individual domains being coded for by separate pieces of genetic material as shown in Figure 5.4. The constant region is the same for all heavy (H) or light (L) chains of a given isotype and is therefore coded for by one *C* gene for an L chain and three *C* genes for an H chain, but the variable region can be coded for by any one of several hundred *V* genes. The lymphocytes have the ability to rearrange these genes during differentiation to produce a large number of different combinations of any *V* gene with the *C* gene. There are also a number of *J* genes (for L chains) and *J* and *D* genes (for H chains), and any one of these can combine with any *V* gene and the particular *C* gene to produce an even larger number of possible combinations. Following this rearrangement within the genes coding for the heavy chain or the light chain, any one H chain can combine with any one L chain to form the antibody molecule. Further diversity, due to somatic mutations, also occurs to increase the possible number of different antibody-binding sites.

Advances in hybridoma techniques and recombinant DNA manipulations have led to the development of novel antibody reagents, sometimes called second-generation monoclonal antibodies. These include molecules such as bispecific antibodies where the two binding sites react with two distinct antigens, chimaeric antibodies which can have the V region from a mouse monoclonal antibody genetically combined with a human C region. However, these types of antibodies are mainly useful for tumour cell diagnosis and immunotherapy, and have little use in the microbiological diagnostic area.

Undoubtedly, the most revolutionary advance in the area since the introduction of hybridoma technology has been the demonstration of expression of heavy chain variable regions (VH) and FV regions (heavy and light chains combined) in *E. coli* (Huse *et al.*, 1989; Ward *et al.*, 1989). The fact that the variable region of the heavy chain could bind to its target molecule without the aid of the CDRs from the light chain had been previously recognized, although the VH fragment binds with a lower affinity than the FV fragment or the whole antibody. Bacterial cells are unable to secrete whole antibody molecules that are functional, but can produce antibody fragments that are smaller and not glycosylated.

Using hybridoma cells as the starting material, mRNA, which is enriched for copies of the antibody genes, is extracted and used as a template to produce cDNA (complementary DNA). The next stage is an amplification step which results in the production of a large number of copies of the DNA that codes only for the variable region, i.e. the $V + D + J$ genes for the heavy chain or the $V + J$ genes for the light chain. This amplification is brought about using a technique called polymerase chain reaction (PCR) (Figure 5.5) where a thermostable enzyme allows the polymerization of DNA between two primers. The

PCR

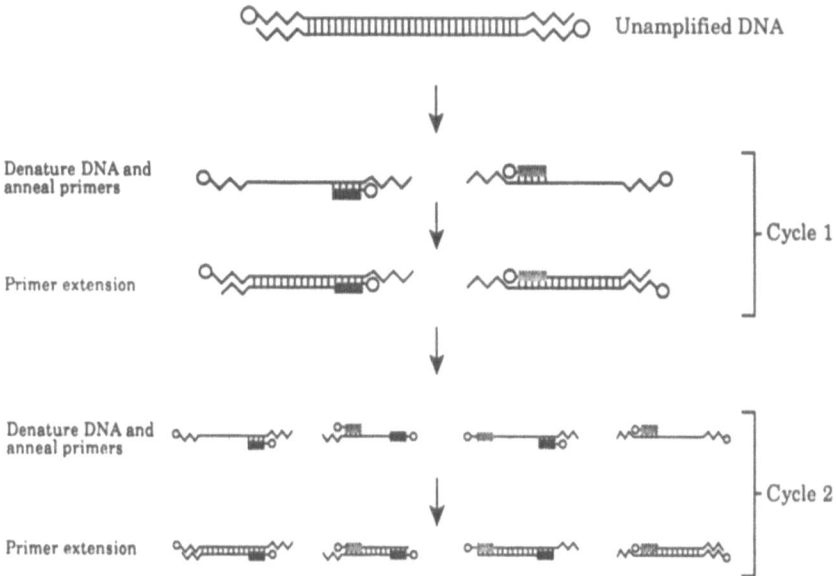

Figure 5.5

The polymerase chain reaction. A thermostable enzyme enables synthesis of the complementary DNA strand from a primer. After 25 cycles the fragment between the primers is amplified approximately 10^7 times

primers, oligonucleotides that are complementary to the terminal stretches of the piece of DNA being copied, correspond to the framework regions of the variable regions of the H and L chains. The framework regions are reasonably constant for all antibody classes; therefore the same set of primers can be used each time, irrespective of the antibody fragment to be amplified.

At this stage the DNA fragment can be spliced into a sequencing vector and the base sequence of the variable region determined. This knowledge is very useful, giving the corresponding amino acid sequence that allows the structure of the variable region (or more importantly the structure of the CDRs, because the framework structure is fairly conserved) to be modelled.

Expression of the fragment relies on the use of an expression vector, which can either be a circular piece of DNA (a plasmid) that will be taken up by the bacteria, or lambda phage DNA which again will be introduced into the bacterial cell. With a plasmid vector the fragment

DNA is inserted, the vector is taken up by the *E. coli* cells, and these are allowed to grow on agar plates. Colonies of cells are picked off the plate and grown in liquid culture. Subsequently, the supernatant can be screened for the secretion of fragments or the cells can be lysed and screened if the fragments have been retained in the periplasm. Lambda phage vectors have the advantage that protein expression can be detected more easily. Phage plaques, which form on an *E. coli* plate, will express the antibody fragment on their surface and can easily be screened for by immunoblotting following transfer of the protein from the plate to nitrocellulose. However, these vectors cannot be used for production of large amounts of the antibody fragment.

Perhaps the greatest potential of the technque is to produce a genomic library of the immunoglobulin repertoire from the spleen cells of an immunized mouse. Using the same techniques as described for hybridoma cells, a large number of VH and VL regions can be expressed separately or in combination. Obviously these artificial combinations of heavy and light chain regions will be different to the combinations produced by the animal, and the selection for high affinity by continuous immunization is lost. However, with this method there is the potential to screen much larger numbers of these randomly associated FV fragments, than there is to screen for whole antibodies in hybridoma production.

Detailed methodology for these techniques is beyond the scope of this book and the reader is referred to the bibliography at the end of the chapter.

In summary, the main advantages of genetically produced antibody fragments are: the speed with which they can be produced, a decrease from several months for a monoclonal antibody to 2 weeks for an antibody fragment; the potential for the much smaller fragment to bind to less well exposed epitopes; and the potential for manipulation of the binding site to alter affinity and cross-reactivity by changing selected bases through site-directed mutagenesis.

—— 5.5 ——

ANALYTE–ANTIBODY BINDING MECHANISMS

The binding of an antibody (Ab) to its analyte (An) can, in general, be considered in the same way as the binding of any ligand to a macromolecule, with the addition of the following characteristics. In an Ab–An reaction the analyte is not irreversibly changed, so the reaction is, in principle, reversible; when considering polyclonal antisera there is enormous heterogeneity of binding reactions with different affinities and different specificities, and antibodies can be raised by design with almost any required specificity. In its simplest terms, the Ab–An reaction can be considered as a reversible bimolecular reaction, described by the following equation:

$$Ab + An \rightleftharpoons AbAn$$

which gives the following definition of the affinity (K_{aff}) of the reaction:

$$K_{aff} = \frac{[AbAn]}{[Ab]\,[An]}$$

where K_{aff} has units of 1 mol^{-1}.

The actual binding site of an antibody consists of residues from the six complementarity determining regions (CDRs) and they form contacts with the binding site on the analyte (epitope); this can be a variety of structures ranging from three to six amino acids of a protein, or three to four residues of a carbohydrate, to a hapten with a molecular weight as low as 100. Binding between these contact residues is due to many weak non-covalent reactions, e.g. hydrogen bonds, Van der Waals reactions, and electrostatic and hydrophobic effects; these numerous weak bonds are the key to the highly specific nature of an Ab–An reaction. Almost as many are repulsive as attractive, and it is this small difference between a large number of attractive and repulsive forces that leads to a net high-affinity bond (a high-affinity bond of 10^{10} 1 mol^{-1} corresponds to the binding energy of only three hydrogen bonds). It is therefore not surprising that a small change in the structure of an epitope can result in an enormous change in the affinity of the reaction. Binding of polyclonal antisera is stable over a pH range of 4–9 and a salt concentration of 0–1.0 mol/l NaCl, therefore conditions such as high or low pH, high salt or denaturing conditions (e.g. 6 M guanidine–HCL or 9 M urea) are necessary to disrupt the bond. However, monoclonal antibodies that do not possess a range of different bonds comprising different types of interactions are more susceptible to changes in pH or salt depending on the actual interaction of the monoclonal antibody with its corresponding epitope. This can be an advantage for immunoaffinity applications because more gentle elution conditions can be employed.

There are two characteristics of an antibody that can be investigated and that will yield a great deal of information about the way the antibody will perform in a certain assay system – the affinity and the specificity. They are not independent of each other, and vary depending on the nature of the assay, emphasizing the importance of characterizing antibodies in the assay system in which they are to be used.

The affinity of an antibody describes the strength with which it binds to the analyte, assuming that one binding site reacts with only one epitope. With a monoclonal antibody there will be only one affinity constant because all of the antibodies react in an identical manner with the analyte; however, with polyclonal antisera the situation is different. Here there are several populations of antibodies that react with different affinities, usually ranging from 10^6 to 10^{10} l mol^{-1}. In most practical situations, the higher-affinity antibodies will dominate, even though they may only be a small fraction of the total antibody population; consequently the effective affinity of a polyclonal preparation is fairly high. Often the converse is true for a monoclonal antibody where the statistical chances of selecting a high-affinity antibody are small, with the result that most monoclonals are of medium affinity. There are several ways to measure the affinity of an antibody for its analyte, the most usual being by equilibrium dialysis or by the use of a radiolabelled ligand. Using the latter the data obtained can be used to plot (bound/ free) against (bound) (referring to antibody bound or unbound to radioactive ligand) which will give $-K_{aff}$ as the slope of the plot. This is a Scatchard plot described by the equation:

$$\frac{B}{F} = -K_{aff}\,B + Ab$$

where B = bound, F = free.

A monoclonal antibody will give a straight line plot but polyclonal antisera will produce a curve (Figure 5.6) which is a reflection of several populations of antibodies with different affinities. The average affinity can be calculated from the slope between the two intercepts. The Scatchard analysis is the most popular way of calculating affinities but there are several other ways. These methods all use data obtained from binding reactions in solutions, i.e. from radio-immunoassays. It is not possible to calculate absolute affinities from ELISA data where one of the reactants is immobilized, but several methods have been published that give relative affinities which, in practice, are almost as useful.

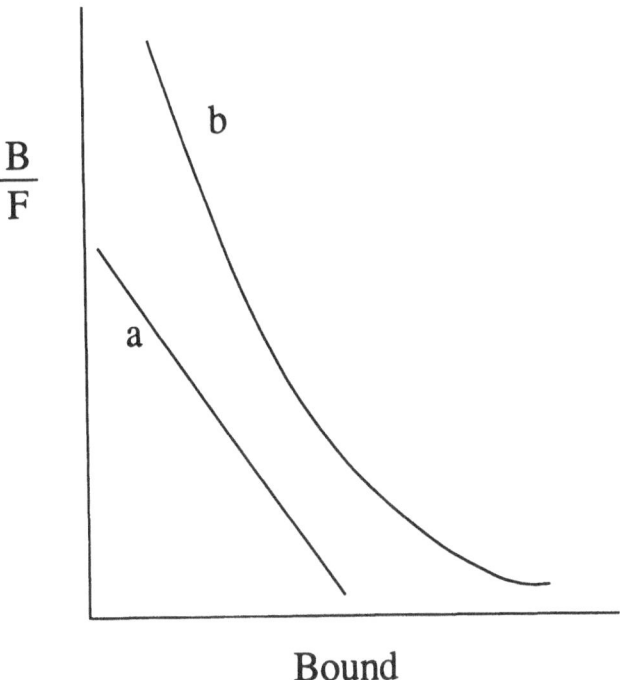

Figure 5.6

Scatchard analysis of (a) a monoclonal antibody, (b) a polyclonal antiserum using radiolabelled ligand. K_{aff} is calculated from the slope of the plots (B = bound and F = free antibody)

The affinity of an antibody relates to one binding site, and should not be confused with the avidity of the antibody. The avidity is the functional affinity which is the result of both binding sites reacting with antigenic determinants and which can, in practice, be very high.

One aspect of an immunoassay which is affected by the affinity of the antibody is its sensitivity. The extent to which sensitivity is affected depends on the assay system (see Section 5.8.1) and its kinetics. In a competitive assay where there is a limited amount of antibody, and both the immobilized and the free analyte are competing for binding sites, the sensitivity is highly dependent on the affinity. However, in an excess-reagent assay, sensitivity is affected minimally by the affinity, but to a much greater extent by the background noise and detection limit of the label.

The other important characteristic of an antibody is its specificity which can be defined as its ability to discriminate between the immunogen and any other molecule not present in the immunogen. Other molecules that are recognized by the antibody, usually due to a structural similarity, are called cross-reactants and will normally react with a lower affinity. If the cross-reactant reacts with a higher affinity than the immunogen, it is said to be heteroclitic.

Basically, cross-reactivity can be divided into two types. Type 1, or true cross-reactivity, is defined as the ability of two analytes to react with the same antibody molecule. This can occur with haptens of very similar structure, or a protein and a peptide with the same amino acid sequence which have lost some tertiary conformation, or even two proteins with a small change in amino acid sequence. In this case complete displacement can be achieved with high levels of the cross-reactant; concentrations higher than that of the analyte are necessary to produce any degree of cross-reaction (Figure 5.7). In the second type of cross-reactivity, shared reactivity, the cross-reactant reacts with a subpopulation of antibodies in a polyclonal preparation. By definition this type 2 cross-reactivity does not occur with monoclonal antibodies. The affinity of the antibody for the cross-reactant may be as great as for the homologous analyte, but displacement reaches a plateau at less than 100% (Figure 5.7). This could occur if the immunogen possessed two epitopes and populations of antibodies were produced to both. If the cross-reactant only possessed one of those epitopes, then it would only react with a fraction of the total antibodies. In cases where this cross-reaction is undesirable, the antiserum can be treated to remove it by passing down an immunoadsorption column using the cross-reactant as the ligand. If the cross-reactivity was type 1 then it could not be removed in this manner because all antibodies would react with the immunoadsorbent, although with a lower affinity. Both types of cross-reactivity can, of course, occur simultaneously with a polyclonal antisera where the cross-reactant only reacts with a certain proportion of the antibodies and with a lower affinity.

Finally, it is worth mentioning that unexpected cross-reactions with apparently unrelated analytes can occur, especially with monoclonal antibodies. An explanation for this is that an antibody binding site can be multispecific and combine with unrelated epitopes in quite different ways. This would occur if there were sufficient numbers of attractive bonds formed between the binding site and epitope, even though they could be an entirely different set of interactions to those that occurred

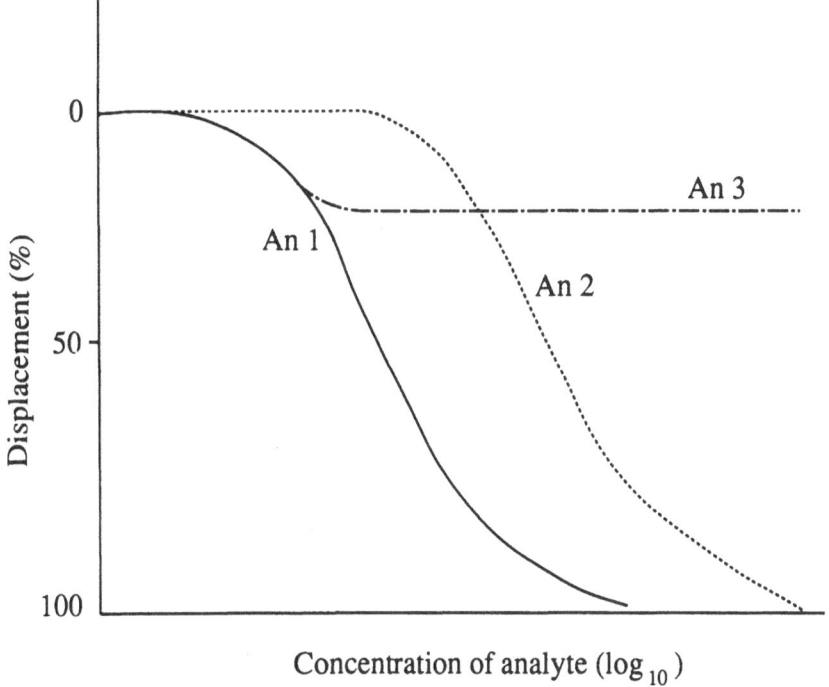

Figure 5.7
Two types of cross-reactivity illustrated in a generalized limited-reagent competitive assay (see Section 5.8.1; An 1 = homologous analyte; An 2 = type 1 cross-reactant; An 3 = type 2 cross-reactant)

with the immunogen. These types of unexpected cross-reactions are seldom seen with polyclonal preparations which consist of many different populations of antibodies. Even if multispecificity occurred with several of these populations, it is highly unlikely that the odd cross-reactions would be with the same molecule for each population. Each odd cross-reaction would therefore only be with a very small percentage of the total antibodies and would, in effect, be diluted out.

—— 5.6 ——
PHASE SEPARATION SYSTEMS

As mentioned earlier, it is necessary to find a way of recognizing that specific binding of antibody and analyte has occurred; the first step in this recognition is usually to separate somehow the bound and unbound phases. Early systems often depended on the high molecular weight of antibodies. Charcoal, which binds low-molecular-weight materials to leave antibody-bound analyte in solution, has been used, and so has ammonium sulphate which has the opposite effect, i.e. it precipitates high-molecular-weight antibody-bound material leaving low-molecular-weight haptens in solution. The traditional microbiological methods of agglutination and precipitation, discussed in Chapter 3, are a form of phase separation.

These methods are very inconvenient to use and do not lend themselves readily to analysis of multiple samples. Almost all separation systems in current use depend on the attachment of one of the components of the assay to a solid surface. If the bound phase of the assay is then subsequently fixed to this surface, the unbound phase can be removed by washing, physical removal of the solid surface or some similar procedure.

The solid surface is generally one of the two following forms: either a bead or particle of some kind, or a continuous surface such as a tube or well in a microtitration plate. The particles can be made of many substances, including agarose, cellulose and latex, with the surface prepared to ensure maximum binding of antibody or analyte. After binding of the analyte to the antibody-coated particle, the particle is then removed from the assay matrix by centrifuging; in some cases, magnetic beads have been used, allowing easier removal by magnetic separation. However, neither method is very convenient, they are not easy to use in automated procedures and washing of the beads is still required. The method of choice is to use a fixed solid surface which can be washed freely to remove unbound material. For multiple assays, this is most frequently the 96-well microtitration plate (Figure 5.8) which can be used in ways ranging from simple manual filling and washing, through semi-automated processes, to fully automated assays using a robotic system such as the Biomek mentioned earlier. These plates readily bind many materials, including antibody and proteinaceous analytes.

For a one-off spot test or for use 'in the field', solid surface formats such as card-tests or dipsticks are being developed; these will be discussed more fully later.

Before finishing this section, a brief mention should be made of non-phase-separation assays, i.e. those in which the reactants are in a homogeneous system. These usually depend on the binding event

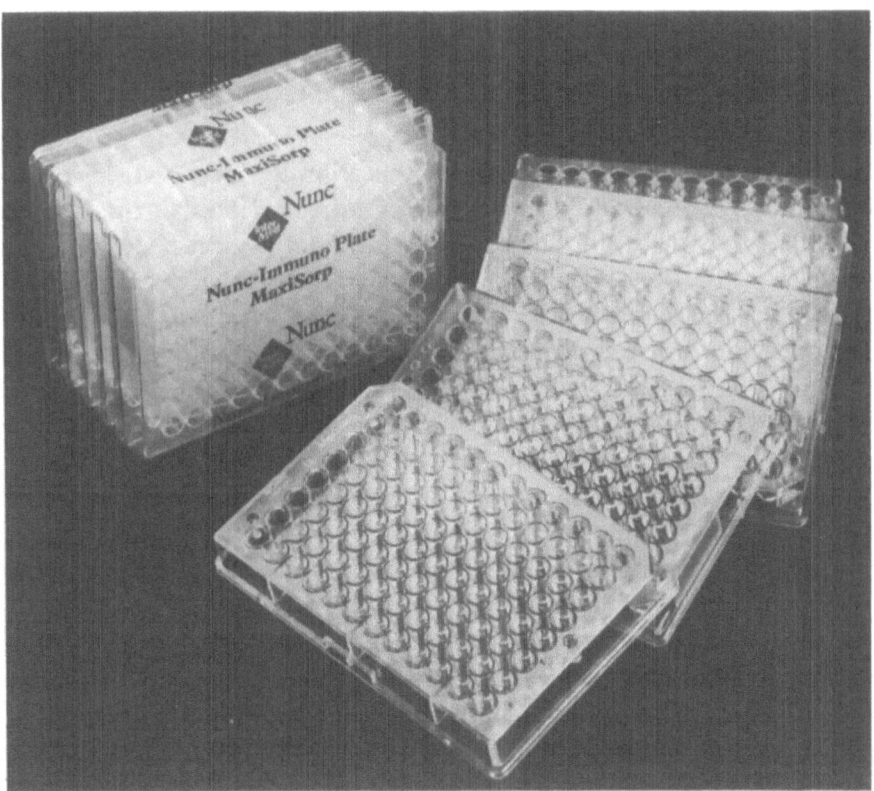

Figure 5.8
Microtitration plates (96-well)

interfering with, and thus reducing, the generation of the signal used for end-point detection (see next section). So far this type of assay has been used mainly for haptens and not applied in the area of food analysis. Although such methods can be very rapid and easy to automate, sensitivity can be comparatively poor and the systems difficult to set up.

—— 5.7 ——
END-POINT DETECTION SYSTEMS

The third of the principal requirements for an immunoassay is a means, preferably quantitative, of recognizing the end-point, i.e. binding of antibody and analyte.

In traditional microbiological use of antibodies, phase separation and end-point detection were simultaneous, with agglutination or precipitation being visible to the naked eye. However, in what would more usually be called an immunoassay the first labels used were radioactive isotopes (^{125}I or ^{3}H), and these assays were termed 'radio-immunoassays' (RIAs). The advantage of radiolabels is that the labelled material is virtually unchanged so that binding is unaffected by the presence of the label. Counts of radioactivity can be made quickly and accurately, but unfortunately labelled materials have a short shelf-life. The other big disadvantage of RIAs is the special handling and disposal facilities needed and, currently, the presence of isotopes in a food laboratory is seen as unacceptable.

The problems of reagent stability and safety are largely overcome by the use of enzymes as labels, and enzyme immunoassays are now the most widely used in the non-clinical field. It has to be realized, however, that the size of the molecule used as a label could change the binding properties of the antibody. The choice of enzyme is fairly wide but it should be stable, have a high specific activity, be easily and cheaply available and be accurately measured; the product is almost always measured by either development of colour or production of light. The enzymes used most often in ELISAs are horseradish peroxidase, alkaline phosphatase, urease and β-galactosidase; the first two are frequently used in commercially available assays. Although some of these enzymes occur naturally in foods and might be believed to interfere in an ELISA, the extensive washing steps between sample addition and colour development eliminate this potential problem. This would, of course, not always be true for the homogeneous assays.

Attachment of the chosen enzyme to the antibody is essentially the preparation of a protein–protein conjugate for which several methods are available. The chemistry of coupling reactions is straightforward, and methods using glutaraldehyde or periodate are most often used; those that have been successful are given in Appendix 6. It is essential that the final preparation is freed of unconjugated material, using standard protein purification techniques.

Substrates suitable for colour development as an end-point have become more widely available. Alkaline phosphatase gives yellow reaction products with *p*-nitrophenyl phosphate (PNP) and two compounds are commonly used as substrates for horseradish peroxidase: 2,2'-azinobis(3-ethylbenzthiazoline) sulphonic acid (ABTS), and 3,3',

5,5'-tetramethyl benzidine (TMB). Details of the use of these enzyme/substrate systems is given in Appendix 5.

As an alternative to colour development, an enzyme label can be used to catalyse a reaction generating light. Luminol is oxidized to produce light by peroxides such as hydrogen peroxide and the reactions proceed much faster in the presence of horseradish peroxidase in an alkaline buffer. This system has not been used extensively in ELISAs for food, and does not seem to offer an advantage over colour development as an end-point.

Production of light from antibody labelled directly with a light-emitting compound, rather than via an enzyme-catalysed reaction, does, however, offer advantages in assay sensitivity. Aryl acridinium esters have been used for this purpose; the light-emitting reaction does not require a catalyst, cleavage of the molecule being achieved by dilute hydrogen peroxide. Unfortunately, unlike the prolonged light output from luminol reactions, the light emission is extremely short-lived. (The rate can be adjusted but with loss of the sensitivity advantage.) It cannot, therefore, be measured with a conventional luminometer but requires an instrument allowing addition of hydrogen peroxide and measurement of light *in situ*. Microtitration plates must be of a special opaque construction. Labelling protocols involving mild conditions, which couple acridinium salts to antibodies via lysine residues, have been developed (Weeks *et al.*, 1983); labelling kits are available (Bioanalysis, Cardiff, UK).

As a variation of the above light-emitting systems, fluorescence detection as an end-point can be used. Again, this can involve indirect or direct labelling of antibody. For the indirect method, alkaline phosphatase- or galactosidase-labelled antibody is used, with a fluorogenic substrate substituted (Appendix 5) for the chromogenic one. With certain ELISA formats these enzymes, using methyl umbelliferyl derivatives as substrates, have shown sensitivity advantages (Ward *et al.*, 1988). However, natural background fluorescence can be a disadvantage, and a fluorescence plate reader, with appropriate excitation and emission wavelengths, would be required for microtitration plate-based ELISAs.

Direct labelling of antibodies with fluorochromes (compounds that fluoresce when excited by light of an appropriate wavelength) was proposed as an early alternative to radiolabelling. However, its current use seems to be limited to two areas: direct microscopic observation of bacterial cells and flow cytometry using a fluorescence-activated cell sorter. Both these methodologies will be covered later in the book. Most fluorochromes now in use are derivatives of fluorescein (green emission), rhodamine or phycoerythrin (both red-orange emission). Use of two antibodies labelled with compounds that give contrasting emission spectra can yield information about differential binding of those antibodies. Conjugation of the isothiocyanate derivatives of fluorescein (FITC) and rhodamine (TRITC) to antibodies via thiourea bonds is straightforward (see Appendix 6).

Although fluorochrome-labelled antibodies allow direct observation of binding to bacterial cells, this is limited by the resolution of the light microscope. If higher magnification studies on bacteria are required, then it is possible to use antibodies labelled with particles of ferritin or gold. These are electron dense and can thus be used as structural probes in electron microscopy of cells.

The last of the commonly used labels for antibody is biotin. This has the advantage that due to its low molecular weight it is less likely to interfere with antibody binding than enzyme labels. However, it has to be used in conjunction with avidin or streptavidin – proteins that have a very high affinity for biotin and that possess four binding sites. If avidin is conjugated with an enzyme, fluorochrome or other end-point detection molecule, then a workable system can be developed. The principal advantage of the avidin/biotin system is that, owing to the four biotin-binding sites on the avidin molecule, a large network of antibody–biotin–avidin–end-point detector can result from a single antibody–analyte binding event. This can lead to considerable enhancement of the assay signal, a feature that will be discussed later in the book.

Conjugation of antibody with biotin is a simple procedure (see Appendix 6), again involving coupling via lysine residues.

During the development of an immunoassay, and sometimes in routine use, it is much more convenient not to have to label the specific antibody used to recognize the analyte, but to use a commercially available pre-labelled antibody. All of the labelling systems discussed can be obtained in 'off-the-shelf' forms that are conjugated to anti-species antibodies, i.e. those that bind to all antibodies derived from a particular animal species. The labelled antibody is used as a second antibody, binding to the first, specific, antibody. Although adding to the complexity of the assay, in addition to the convenience it may allow use of a specific antibody the binding of which would otherwise be disturbed by being labelled directly.

—— 5.8 ——
IMMUNOASSAY FORMATS

5.8.1 ELISA

The enzyme-linked immunosorbent assay based on a 96-well microtitration plate is the most widely used immunoassay format. It is convenient to discuss basic assay systems in terms of a plate ELISA, but often they will apply to many other immunoassay formats.

Although many configurations of ELISA have been devised, most are variations of two basic methods: the limited-reagent and excess-reagent systems. Additionally, either analyte or antibody can be labelled; in the former case, a quantity of labelled analyte is introduced into the system as a means of determining the unknown (unlabelled) analyte in the sample. Alternatively, unlabelled analyte can be introduced but in a different phase from the unknown sample analyte as a coating on the solid surface. In either case, the added analyte (labelled or unlabelled) can be distinguished from the sample analyte.

In a limited-reagent system, the quantity of the added labelled reagent (whether analyte or antibody) is pitched at a level at which the number of binding events involving the added reagent will be less than the potential maximum, i.e. the quantity of one partner exceeds that of the other; the remaining unbound partner is therefore able to bind to the unlabelled equivalent. This type of reaction is also frequently known as a 'competitive' assay, due to the supposed competition for binding between introduced analyte and sample analyte. At equilibrium, a distribution of binding exists that is proportional to the concentration of analyte in the sample. Whether the labelled reagent is analyte or antibody, the higher the sample analyte concentration, the less label will be present at the end of the assay; the amount of product (colour or light) is inversely proportional to analyte concentration. The lowered binding of antibody to the solid phase caused by the presence of analyte in the sample is termed 'displacement'. Figure 5.9 illustrates a limited-reagent system using labelled antibody.

In excess-reagent assays, labelled antibody is used. (Wherever reference is made to labelled antibody, it should be noted that instead

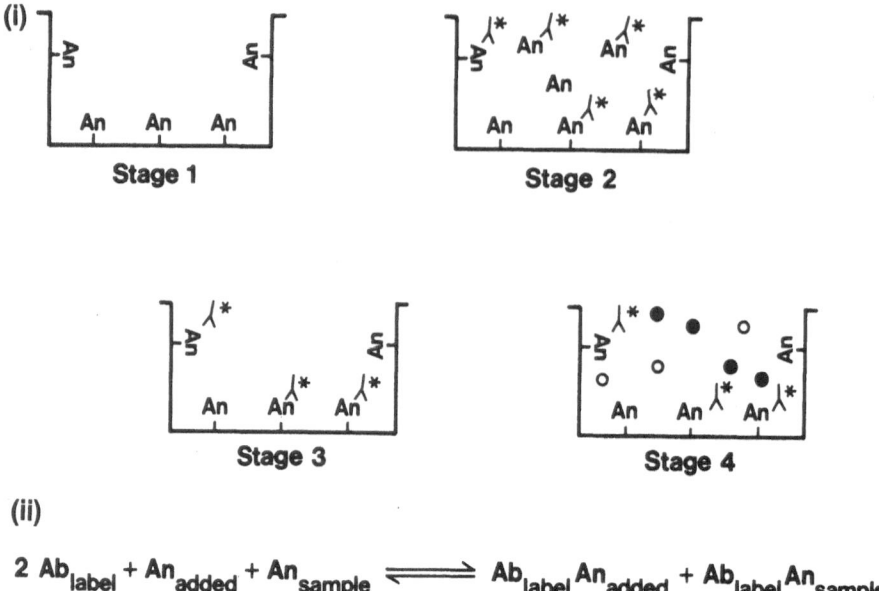

(ii)

$$2\,Ab_{label} + An_{added} + An_{sample} \rightleftharpoons Ab_{label}An_{added} + Ab_{label}An_{sample}$$

Figure 5.9

Principle of a limited reagent competitive assay using a solid-surface separation technique. (i) Stage 1: solid surface pre-coated with a preparation of the analyte. Stage 2: simultaneous incubation with sample (containing the analyte to be measured) and a limiting concentration of enzyme-labelled antibody. Stage 3: antibody bound to sample analyte, and unbound material, both removed by washing. Stage 4: addition of substrate and colour development, inversely proportional to concentration of analyte in sample. (ii) Generalized equation for this assay (λ * = labelled antibody; Ab_{label} = labelled antibody; An = analyte; An_{added} = analyte pre-coated to solid surface; An_{sample} = unknown analyte in sample; $\circ$ $\bullet$ = substrate colour development)

of a labelled *specific* antibody, a second labelled *anti-species* antibody could usually be used instead.) Sufficient unlabelled antibody (in practice, an excess) to capture all the analyte in a sample is bound to the solid surface (the 'capture' antibody). After capture and further washing, labelled antibody (the 'detector' antibody) is added which binds to the captured analyte (this assumes that the analyte has more than one epitope available) to form an antibody–analyte–labelled antibody 'sandwich' (Figure 5.10). If a (third) labelled anti-species antibody is to be used to recognize an unlabelled detector antibody, then the detector must be from a different animal species to the first, otherwise the anti-species antibody will bind to the capture antibody coating the plate. Alternatively, the capture antibody can be reduced to its Fab fragments (see Section 5.2) by treatment with pepsin (Adams

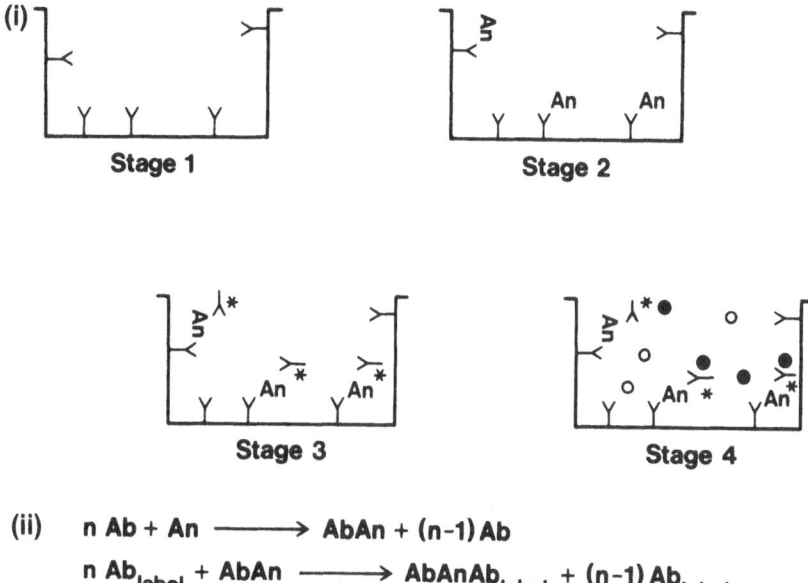

(ii)   n Ab + An $\longrightarrow$ AbAn + (n-1) Ab

         n Ab$_{label}$ + AbAn $\longrightarrow$ AbAnAb$_{label}$ + (n-1) Ab$_{label}$

Figure 5.10

Principle of an excess reagent sandwich assay using a solid-surface separation technique. (i) Stage 1: solid surface pre-coated with an excess of the capture antibody. Stage 2: incubation with sample containing the analyte to be measured. Stage 3: incubation with an excess of enzyme-labelled detector antibody. Stage 4: addition of substrate and colour development, directly proportional to concentration of analyte in sample. (ii) Generalized equations for this assay (for key see Figure 5.9)

and Barbara, 1982); the detector antibody is then recognized by addition of enzyme-labelled protein A, which will bind only to the intact Fc region.

In a slight variation of the sandwich assay, a detector antibody is used which binds to a different epitope from the capture antibody; this can increase the specificity of the assay as the chances of unwanted cross-reaction are reduced due to a lower probability that a potential cross-reactant will possess two epitopes identical to those on the target analyte. This type is known as a 'two-site' assay. In sandwich-type assays, the amount of label present at the end of the assay will be directly proportional to the concentration of sample analyte, and thus the amount of product (colour or light) will increase with analyte concentration.

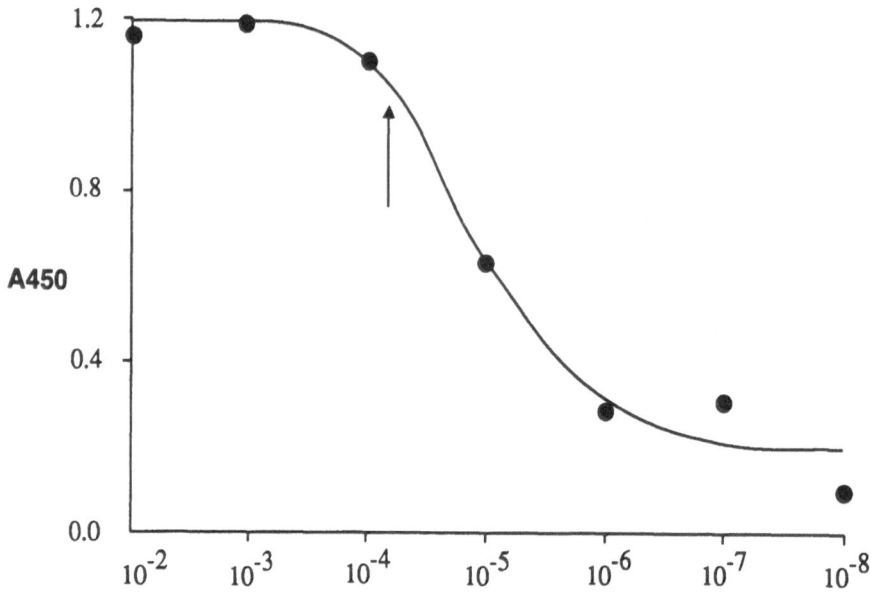

Dilution of antiserum

Figure 5.11
Typical titration curve for a polyclonal antiserum (raised in a rabbit to Salmonella
enteritidis; *the solid surface of a microtitration plate was coated with cells of* S. enteritidis,
*1 µg protein/ml, and unoccupied binding sites blocked with bovine serum albumin, 10 g l^{-1};
binding of rabbit antibodies was detected with horseradish peroxidase labelled anti-species
antibody and the substrate was tetramethylbenzidine, stopped and absorbance read at 450 nm.
The initial choice of antibody concentration for a limited-reagent assay should be made from the
arrowed region)*

Simple antiserum titres will usually be the first analyses performed
during development of an assay system. A microtitration plate is coated
(see Appendix 7) with the analyte of interest and wells loaded with
replicate volumes from a dilution series of the serum made up in buffer
(see Appendix 7). After an incubation period at 37°C (2 h is sufficient)
and a washing step (see Appendix 7), appropriate anti-species, enzyme-
labelled antibody is added; after further incubation and washing steps,
substrate is added and colour development measured. A typical titration
curve is shown in Figure 5.11.

The simplest first assay for detection of analyte in a sample is a
limited-reagent (or competitive) assay, with the added analyte and

unknown sample analyte distinguished by being in separate phases. Detection of sample analyte is by specific antibody and anti-species-labelled antibody. Microtitration plates are coated with a sample of analyte as for titration curves. A limiting concentration of specific antibody is chosen by reference to a titration curve performed on that antibody; a starting point for determination of the optimum concentration is to use a dilution giving just less than maximum optical density in the titration curve (Figure 5.11). Maximum sensitivity in the assay may require a different concentration of antibody and this can be determined by further experimentation. Equal volumes of sample and chosen antibody concentration are incubated simultaneously in wells of the coated plate; after a washing step, labelled antibody is added as for the titration curve. Further incubation and washing are followed by substrate addition. For a typical assay protocol see Appendix 7. Figure 5.12 shows an example of a standard curve for a limited-reagent assay.

For a large analyte such as a bacterial cell or protein toxin, sandwich ELISAs involving two antibody-binding sites on the analyte can be developed. The usual format is to use a polyclonal antibody as a plate coating to capture the cell, and a monoclonal antibody as the detector antibody forming the upper layer of the 'sandwich' (although other groups reverse this arrangement). The monoclonal antibody can itself be labelled with enzyme or other suitable label, or an anti-species labelled antibody can be used to detect the monoclonal antibody. The polyclonal antibody can, and usually does, have a broader specificity than the monoclonal antibody and its higher affinity will ensure maximum capture of analyte from the sample. Normally, a rabbit polyclonal and a mouse monoclonal antibody would be used. A protocol for a typical sandwich-type ELISA is shown in Appendix 7 and a standard curve is shown in Figure 5.13. During development of an assay, the concentration of the antibodies used can be varied to test the effects on assay performance, but these concentrations are less critical in an excess-reagent assay than in a limited-reagent system.

When developing an assay, and probably also while assessing a commercially available assay, it is necessary to determine the cross-reactivity, both desirable and undesirable. It is helpful to know the degree of any cross-reactivity rather than just a positive or negative result, and this is usually done by comparison of standard curves of the principal target and any other potential target to be assessed. It is possibly merely to perform antibody titrations on plates coated with the

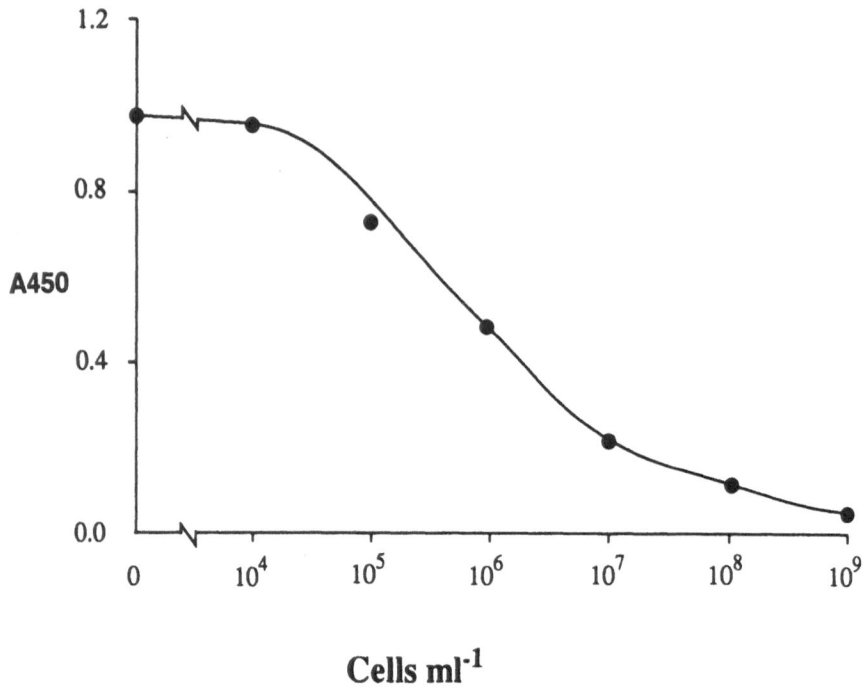

Figure 5.12
Typical standard curve for a limited-reagent competitive assay (analyte is S. enteritidis; *the solid surface of a microtitration plate was coated with cells of* S. enteritidis, *1 μg protein/ml, and blocked with bovine serum albumin 10 g* l^{-1}; *the specific antibody was a rabbit polyclonal serum diluted 1:20 000; binding of rabbit antibodies was detected as for Figure 5.11)*

potential analytes, but because the binding kinetics may differ from those in the final assay format, it is best to use the complete assay for this purpose. Figure 5.14 illustrates an example of the comparison of standard curves and calculation of the extent of cross-reactivity.

In general, excess-reagent assays are more sensitive than limited-reagent types, mainly because the excess antibody can bind all of the analyte present in a sample; where two different antibodies recognizing different epitopes are used, then they can also be more specific. However, they are wasteful of antibody and cannot be used for haptenic molecules possessing only one copy of the epitope; such molecules are not at a disadvantage in the detection of bacterial cells because the latter usually have many sites for the epitope.

Several ELISAs are available commercially for salmonellae and *Listeria* spp. in foods and examples will be discussed in Chapter 8.

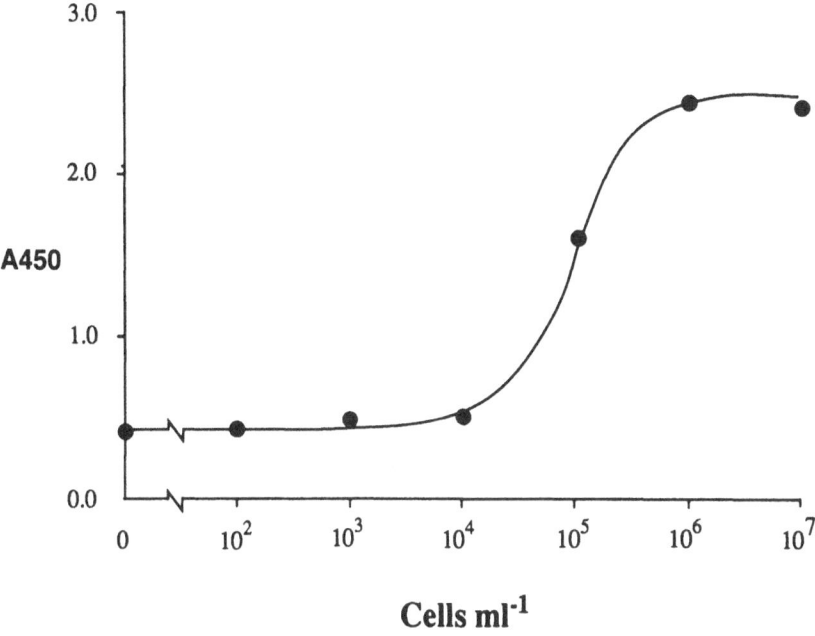

Figure 5.13

Typical standard curve for an excess-reagent sandwich assay. Analyte is S. typhimurium; the solid surface of a microtitration plate was coated with a rabbit polyclonal serum, diluted 1:1000, as the capture antibody and blocked with bovine serum albumin 10 g l^{-1}; the detector antibody was a mouse monoclonal antibody culture supernatant, diluted 1:20; binding of the detector antibody was quantified as for Figure 5.11

Many novel variations of the basic immunoassay formats have been developed in recent years, each with its own carefully contrived acronym, particularly in the field of homogeneous (non-separation) assays. These assays, based mostly on the principle of the binding of antibody to target changing the activity of the enzyme label in solution (Figure 5.15), have all been developed for haptens and have, as yet, made no contribution to the detection of bacteria.

The two basic ELISA formats have now been thoroughly tested and novel formats, which usually increase the complexity of the system, could easily lead to a reduction in robustness of the assay. It would be unfortunate if this led to a loss of confidence in immunoassays; in the authors' opinion, simplicity in use is one of the main advantages of the technique and one of the most attractive to a potential user.

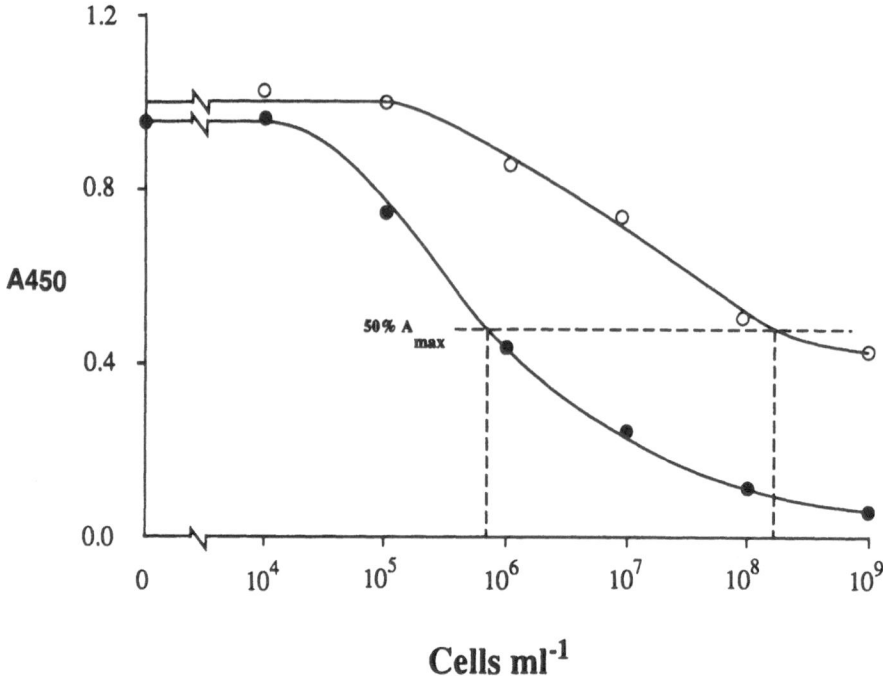

Cells ml^{-1}

Figure 5.14
Determination of cross-reactivity from assay standard curves (● = S. enteritidis, ○ = Citrobacter freundii; limited-reagent assay with conditions as for Figure 5.12). Cell concentrations at 50% maximum absorbance (i.e. 50% displacement from zero) are compared to give a measure of cross-reactivity:

$$\text{Percentage cross-reactivity} = \frac{\text{Concentration of analyte}}{\text{Concentration of cross-reactant}} \times 100$$

$$\text{for C. freundii} \quad \frac{7.0 \times 10^5 \text{ cells ml}^{-1}}{1.6 \times 10^8 \text{ cells ml}^{-1}} \times 100$$

$$= 0.004\%$$

5.8.2 Latex agglutination methods

As mentioned earlier, one of the traditional uses of antibodies in microbiology was in the serotyping of pure cultures by slide agglutination tests which can be read by eye or microscopically. This type of reaction, where the antigen is particulate (i.e. a bacterial cell) and the antibody in solution, is a standard agglutination. This format, although still widely used for serotyping bacteria, has not extended beyond this particular use. However, a format termed 'reversed passive

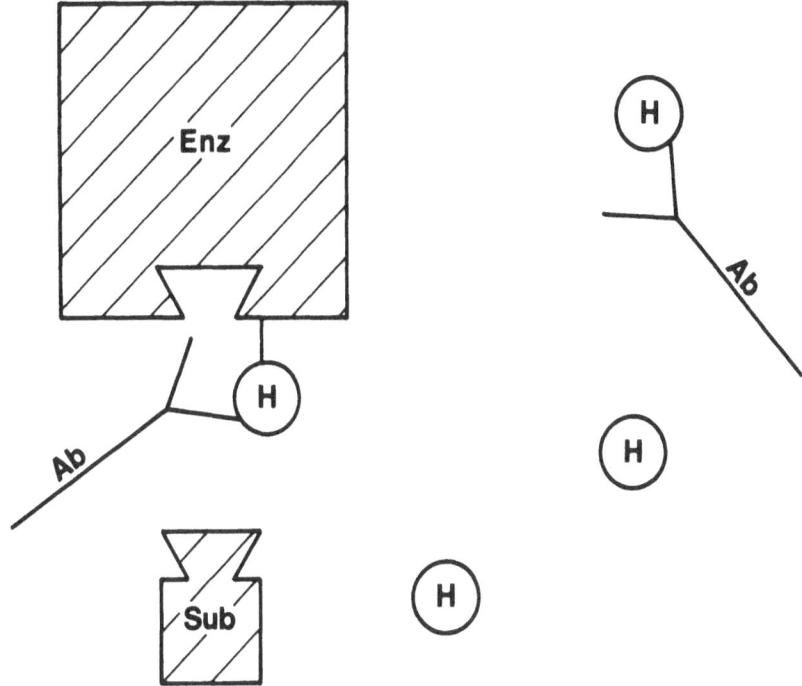

Figure 5.15
One form of homogeneous enzyme immunoassay (Enz = enzyme; H = hapten; Ab = antibody; Sub = substrate) in which binding of antibody to enzyme-labelled hapten physically prevents access of substrate to enzyme. The degree of this inhibition reflects the concentration of free hapten in the reaction mixture

agglutination' has been developed further. This agglutination is termed 'reversed' because the antibody is bound initially to particles, and 'passive' because these particles play no active part in the reaction, merely acting as a carrier. In particular, sensitive agglutination assays for bacterial toxins using antibody bound to polystyrene latex particles are available commercially and one is discussed in Chapter 8.

One useful property of latex particles is that they can be produced in a coloured form, so that when in non-agglutinated suspension the colour is diffuse, but when aggregated a definite colour is discerned by eye. In this way an initial mixture of differently coloured particles, each colour being coated with a particular antibody, can be used to give differential colour reactions depending on the antigen present in the sample. A system for grouping salmonellae based on this principle is available commercially.

5.8.3   Microscopy using labelled antibodies

Use of fluorescently labelled antibodies (FA) was discussed earlier in the book and FA detection of salmonellae in food has received 'First Action' approval from the Association of Official Analytical Chemists in the USA (Anon, 1975). However, the method would seem to be at a severe disadvantage for the routine detection of pathogens in food compared to ELISAs, because it is lengthy and difficult to automate. Although attempts have been made in this direction (Barrell and Paton, 1982) at least 10^5 bacteria/ml are needed for even one cell on average to be visible per microscope field.

Nevertheless, the method may have applications in non-routine use where it is desirable to determine the location of bacterial growth (Dodd, 1990) in a non-homogeneous food sample. The formation of microcolonies associated with particular food components could be visualized in suitably stained sections. An epifluorescent microscope is required.

Similarly, electron microscopy can be carried out using gold-labelled antibodies, although the high magnification available would not be necessary for visualization of cells or microcolonies in food.

5.8.4   Flow cytometry

At its most basic, a flow cytometer (FC) could be regarded as an automated particle detector and counter. However, in addition to this basic function, an FC can give extra information about the nature of that particle; this information can be intrinsic, e.g. the particle size, or extrinsic, such as whether or not a particular stain has been taken up. If a suitably labelled antibody is used (usually a fluorescent label), then binding of this antibody to the particle can be recognized. The principles of an FC are shown in Figure 5.16.

The fact that each particle is examined individually at high speed in a continuous flow seems to indicate that the instrument would be suitable for detection of bacteria in liquid systems. Although commonly used for analysis of eukaryotic cells, extension to prokaryotes is a recent development; the size of bacteria approaches the limit of sensitivity of most instruments. Work on a system for detection of pathogens in food, employing the dual parameters of cell size and binding of specific fluorescent antibody, is in progress. Initial experiments, using cell size and nucleic acid staining as measurement parameters, have shown very promising limits of detection in the order of 10^2 cells/ml for bacteria in pure culture (Pinder et al., 1990).

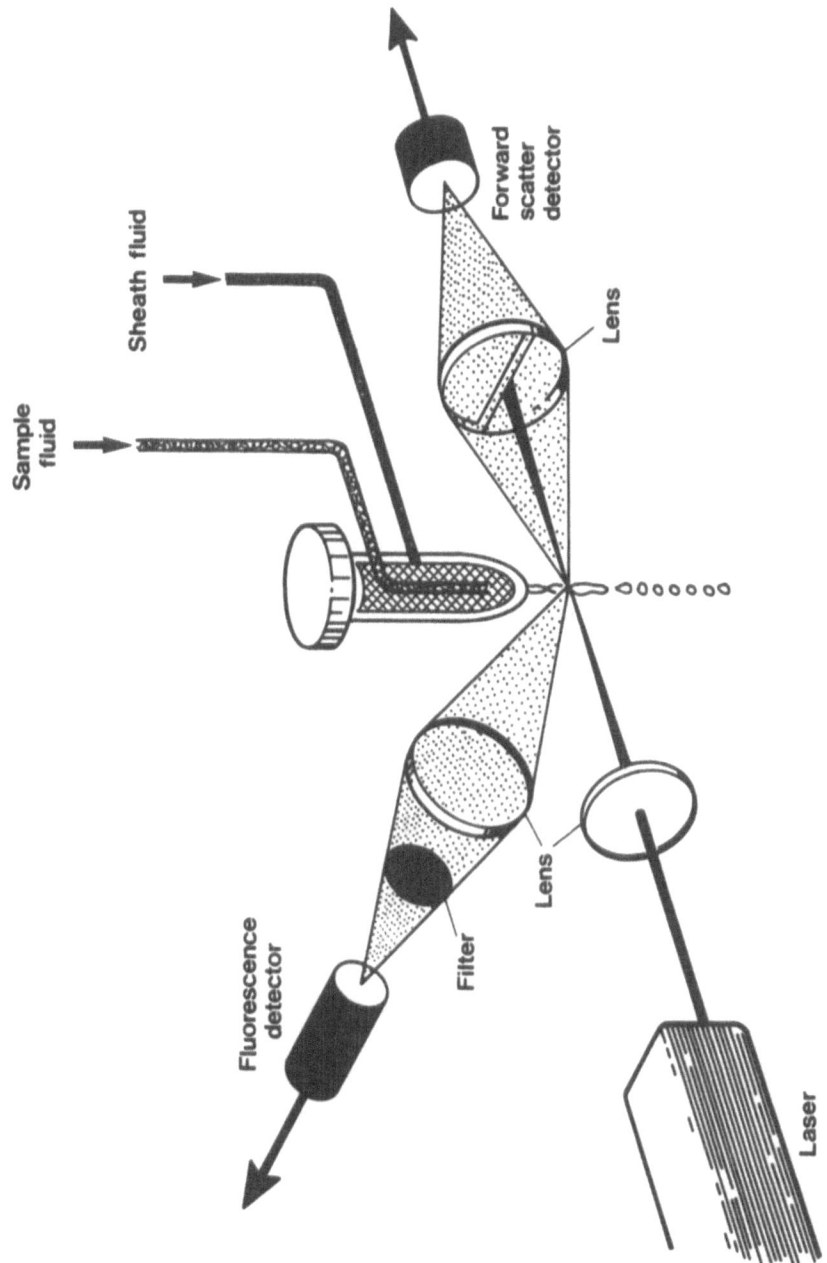

Figure 5.16
Principles of a flow cytometer for fluorescent antibody detection of bacteria (based on Pinder et
al., *1990, with permission)*

The nature of the detection method is such that a number of events need to be plotted to be sure that a cluster of these events represents a definite population of bacterial cells, and not just random particles (Pinder et al., 1990). The ideal situation for this type of instrument would be where a fluid preparation is to be sampled on-line in real time, for example, in a sidestream diverted from a processing line. The computing capabilities associated with the instrument could record the passage of organisms and accumulate the data until some threshold had been reached, whereupon an automatic warning could be given.

5.8.5 Cell concentration methods

A principal advantage of antibody-based methods is the ability to recognize the target cell among a diverse background of non-target organisms and food components. An alternative approach to the detection of pathogens is to remove the organisms physically from that environment through the use of antibodies.

Several methods can be used for this cell separation and concentration process (Wyatt et al., 1991). Immunoaffinity columns, in which the antibody is bound to agarose beads, can be constructed. Coupling of antibody to agarose gel is straightforward if gel activated by cyanogen bromide treatment is used (see Appendix 8); alternatively, protein-A-linked gels are available which can bind antibody via the Fc portion of the molecule, leaving the specific binding sites free. Large volumes of liquid containing small numbers of organisms can be passed through the column, concentrating the cells in the antibody-coated gel. Slurries of food can be made in buffer, clarified by low-speed centrifugation and applied to the column. After washing away unbound material, captured cells can be eluted and measured by ELISAs or other specific methods. Incidentally, although outside the scope of this book, this principle is successfully used in a commercial assay for detection of aflatoxins in food and could be used for concentration of bacterial toxins in a similar manner.

An alternative method to separate cells is to use a solid surface similar to the use of a microtitration plate for an ELISA. If a sterile petri dish is coated with antibody, live cells can be captured by incubating the food sample as a slurry in the dish for a short period. After capture, unbound material is washed away and a layer of cooled but still molten agar medium added to the dish. After an incubation stage (overnight or longer, depending on the organism) colonies develop in the agar where cells were captured on the petri dish surface. If a diagnostic agar is used, then colony characteristics offer assurance of correct identification of the organism additional to that given by the use of specific antibodies for capture. Cells can also be specifically captured from a mixed flora using other solid surfaces such as dipsticks. The captured cells are then washed free of unbound material, transferred to a liquid medium for outgrowth and ultimately detected by a specific ELISA. Work on these immunocapture and immunoenrichment techniques is at an early stage of development (Wyatt et al., 1991).

Methods based on electrical changes (conductance, impedance) produced during growth of organisms in culture are well established, but are generally useful only for total number of organisms in a food. It has been suggested that the limited applicability to specific detection of pathogens could be overcome by the use of antibodies to capture target cells live from a mixed flora, prior to transfer to the electrical system. This system, termed 'immunomodified electrical technique' is under development for salmonellae (Bird et al., 1989). Figure 5.17 illustrates the principles of immunocapture methods.

5.8.6 Immunosensors

The term 'immunosensor' has been applied to at least two rather distinct systems. A true immunosensor could be defined as a device that gives a direct change in an output signal of an electrical, optical or acoustic nature as a result of the antibody–antigen binding event. Such devices are at a very early stage of development and are an extension of enzyme electrode technology, involving immobilization of antibodies on a membrane or other surface (Figure 5.18). As with the flow cytometer, these instruments would seem to offer the possibility of true on-line monitoring in the future, given sufficient sensitivity.

The other type of immunosensor is one in which the signal is generated as a result of using a labelled antibody, in an otherwise conventional ELISA. For example (Mirhabibollahi et al., 1990), catalase-labelled antibodies have been used to release oxygen from hydrogen peroxide in an assay for *Staphylococcus aureus* in food; the oxygen output is then measured with an amperometric electrode. At present, this system is really only an alternative end-point detection method, offering little advantage over colorimetric or light-based end-points.

5.8.7 Signal amplification in ELISAs

The use of an enzyme/substrate system has an in-built amplification compared to direct labelling with radio-isotopes, in that catalysis of

i)

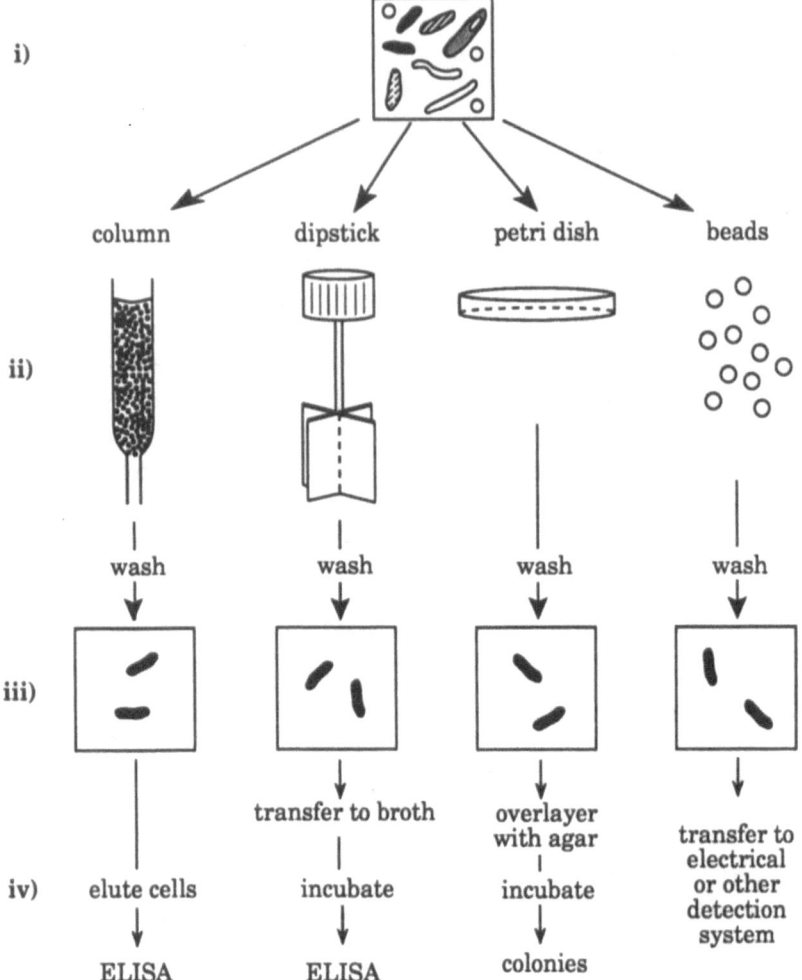

Figure 5.17
Outline of some possible capture, enrichment and concentration methods for bacterial pathogens using antibodies. (i) Initial mixed population, including pathogen, (ii) specific binding of pathogen to antibody-coated surface, (iii) other organisms removed by washing, leaving bound pathogen, (iv) application of detection system

product from substrate continues with time. There are, however, several ways of improving the level of signal produced and hence, in theory, the sensitivity of the assay. For reasons discussed in Section 5.9, it will not always be necessary to push for maximum sensitivity in an

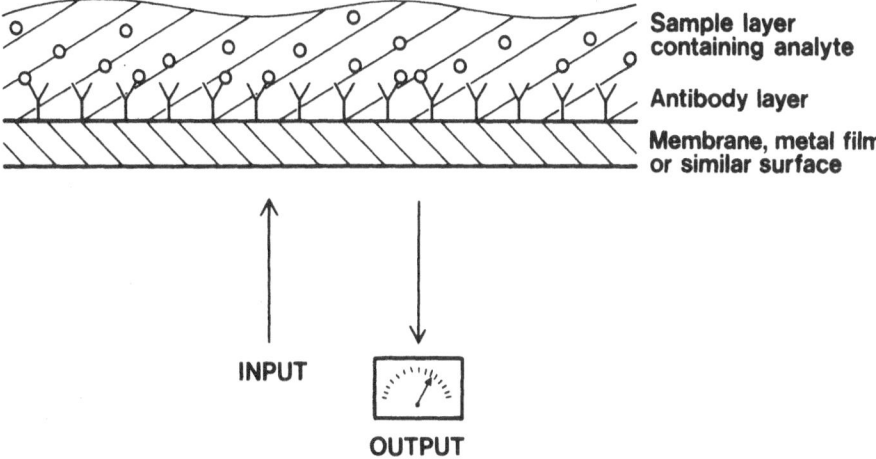

INPUT

OUTPUT

Figure 5.18
Generalized diagram of an immunosensor. The input could be optical, acoustic or electrical; the
output signal depends on the degree of binding of analyte to immobilized antibody

ELISA for bacterial pathogens in food; however, for detection of pre-
formed toxins, the lower the detectable level the better.

The first and most obvious way of increasing the signal is, of
course, as indicated above, i.e. longer incubation times, both to
increase the degree of binding by labelled antibody and to allow
maximum conversion of substrate to product. This, however,
immediately removes one of the main advantages of immunoassays,
their speed; additionally, it may increase the assay background.

A second approach is to increase the degree of labelling relative to a
single antigen–antibody coupling. The most widely tested way of doing
this is to use the avidin–biotin system. Avidin (an egg white protein)
and streptavidin (produced from a *Streptomyces* species) have four high-
affinity binding sites per molecule for the small vitamin biotin (see
Section 5.7). If a specific antibody (or a secondary, anti-species
antibody) is labelled with biotin, and an enzyme or other detector
molecule is labelled similarly, then avidin can be used as a bridge
between antibody and enzyme, giving a 3:1 ratio of enzyme to antigen
(Figure 5.19). In a further refinement of this system, the enzyme is
labelled with more than one biotin molecule enabling a large network
to develop (Figure 5.19) with a much higher enzyme:antigen ratio.

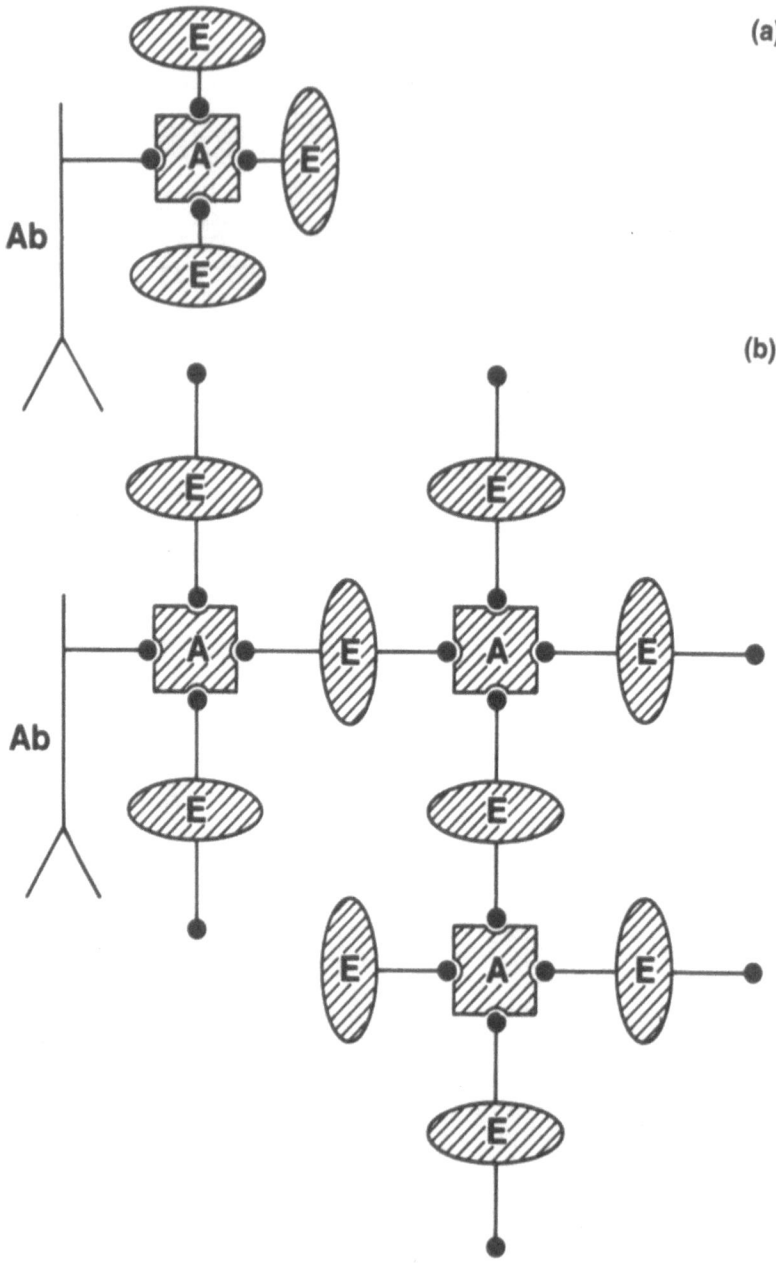

Figure 5.19
Diagrammatic representation of avidin–biotin amplification system. (A = avidin; Ab = biotin-labelled antibody; E = biotin-labelled enzyme). (a) Avidin as a bridge connecting antibody to enzyme in a ratio of approximately 1:3; (b) extension of the system using multilabelled enzyme, giving greater ratios of enzyme to antibody

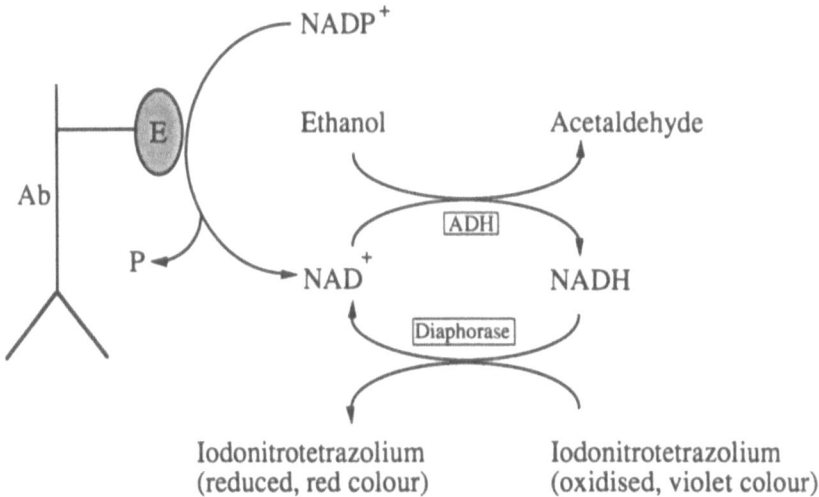

NADP⁺

Ethanol   Acetaldehyde

Ab

E

ADH

P   NAD⁺   NADH

Diaphorase

Iodonitrotetrazolium
(reduced, red colour)

Iodonitrotetrazolium
(oxidised, violet colour)

Figure 5.20
Enzyme cycling amplification system. (Ab = antibody; E = alkaline phosphatase enzyme label;
ADH = alcohol dehydrogenase.) The red product is measured by absorbance at 490 nm

These systems are available commercially (Amersham International plc, Amersham, UK).

A third way of enhancing signal is to change the end-point detector. Mention has already been made (see Section 5.7) of possible increases in sensitivity which can be achieved with fluorometric methods and direct labelling with light-emitting acridinium esters. Merely changing a colorimetric substrate may help as some products have a higher molar absorbance coefficient than others.

The final method to be mentioned is a patented enzyme cycling system, marketed by IQ Bio (Cambridge, UK). Antibody labelled with alkaline phosphatase is used in a normal assay, but the enzyme is then allowed to catalyse the conversion of $NADP^+$ to NAD^+. This is then fed into a cyclic system involving two further enzymes and a coloured product (Figure 5.20). A typical 100-fold increase in absorbance is claimed.

— 5.9 —
APPLICATION TO FOOD SAMPLES

A common standard for detection of pathogens in food is that the organism should be absent from a 25 g sample. (The way in which a representative 25 g sample can be taken from a larger batch of product is, of course, a subject in itself.) As discussed in Chapters 3 and 4, many foods will also contain other bacteria which may, in some circumstances, be sublethally damaged. Thus, the analyst is faced with the problems of detecting an organism which might be present in low numbers, outnumbered by closely related species and which is damaged. Each of these factors needs to be addressed if a system using antibodies is to be successful.

The central concept in most protocols for pathogen detection (Figure 5.21) is the use of a selective culture medium; the problems introduced by this step are outlined in Chapter 3. Given the lack of complete specificity of the selective compounds, a better method of differentiating the pathogen from closely related organisms would be to use specific antibodies, and a suitable ELISA would attend to this factor. However, direct detection of a single organism in a 25 g food sample is a much more difficult problem. Even if sufficient signal amplification could be achieved from that single cell, the problem would still remain of finding it in the sample; the old epigram about a needle in a haystack seems to apply here. The approach taken in the authors' experimental work (Lee *et al.*, 1989) is still to include a cultural step in the protocol (Figure 5.22). This is partly to allow damaged cells to recover full competence, which is the function of the pre-enrichment step in the conventional protocol. Additionally, however, the food is thus diluted, lessening any matrix interference (see Chapter 7) and the very considerable amplification achieved by growth of the bacterial cell is exploited – the needle enlarges to be much nearer to the size of the haystack. This is the reason behind the earlier statement that it is not always necessary to develop maximum sensitivity in the ELISA. It has been shown (Lee *et al.*, 1990) that, using specific antibodies in an unamplified ELISA, it is possible to detect salmonellae in this way, even when initially vastly out-numbered.

The above approach, although attractive, has not yet been fully proven in all situations. Commercially available ELISAs (see Chapter 8) for pathogens still maintain a considerable proportion of traditional selective culture protocols; in addition, as another side effect of the use of chemically selective cultures, a 'post-enrichment' step is often added to ensure full expression of surface antigens, which is not usually achieved in the presence of toxic compounds. The replacement of these

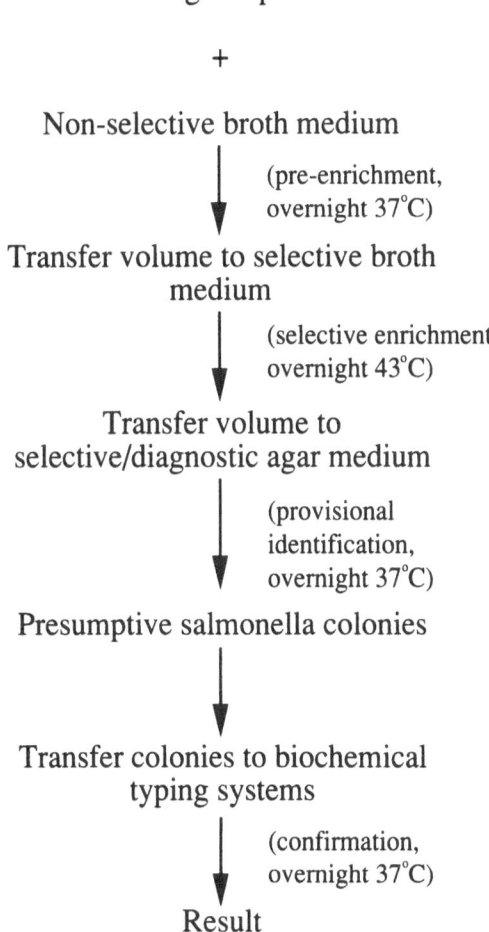

25-g sample

+

Non-selective broth medium

(pre-enrichment,
overnight 37°C)

Transfer volume to selective broth
medium

(selective enrichment,
overnight 43°C)

Transfer volume to
selective/diagnostic agar medium

(provisional
identification,
overnight 37°C)

Presumptive salmonella colonies

Transfer colonies to biochemical
typing systems

(confirmation,
overnight 37°C)

Result

Figure 5.21
Selective enrichment culture procedure for pathogenic bacteria in foods, illustrated by a protocol for detection of salmonellae

compounds by specific antibodies would clearly be a major step forward in pathogen detection.

All of the bacterial toxins commonly formed in food are water soluble and thus a relatively simple extraction procedure can be used. Food samples (usually 10 g) are homogenized in saline or suitable buffer and clarified by centrifugation and/or filtration. The filtrate is then used directly in the immunoassay. If greater sensitivity is required, then

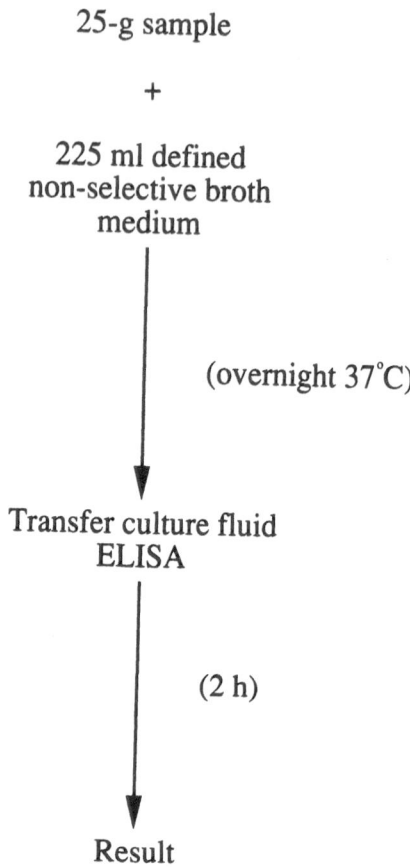

25-g sample

+

225 ml defined
non-selective broth
medium

(overnight 37°C)

Transfer culture fluid
ELISA

(2 h)

Result

Figure 5.22
*Protocol for detection of salmonellae in foods developed at the AFRC Institute of Food Research,
Norwich, UK*

larger amounts of food can be extracted and the filtrate concentrated by
one of a number of common techniques, such as ultrafiltration. The
assay system should be checked for interference from any remaining
food matrix – this is dealt with in Sections 7.1 and 7.2. Application of
immunoassays to detection of toxins in foods has recently been reviewed
by Noterman and Wernars (1991).

BIBLIOGRAPHY

Development and Application of Immunoassay for Food Analysis (1990) Ed. J. H. Rittenburg, Elsevier Applied Science, London, UK

Immunoassays in Food Analysis (1985) Eds B. A. Morris and M. N. Clifford, Elsevier Applied Science, London, UK

Immunoassays for Veterinary and Food Analysis (1988) Eds B. A. Morris, M. N. Clifford and R. Jackman, Elsevier Applied Science, London, UK

Monoclonal Antibodies: Principles and Practice (1986) J. W. Goding, Academic Press, London, UK

Fundamental Immunology (1984) Ed. W. E. Paul, Raven Press, New York, USA

Molecular Cloning – A Laboratory Manual (1987) T. Maniatis, E. F. Fritsch and J. Sambrook, Cold Spring Harbour Laboratory, New York, USA

—6—

Equipment and Instrumentation

As the microtitration plate ELISA is by far the most common immunoassay in current use, the equipment discussed in this chapter will be largely for this format. Several other types of analyses likely to be required in a food analyst's laboratory, e.g. for mycotoxins, meat speciation, gluten and soya proteins, are available in this format so any equipment purchased is liable to be of wider use than simply for microbiological assays.

—— 6.1 ——
EQUIPMENT AVAILABLE

The microtitration plate-based ELISA is a very flexible format which allows handling in a variety of modes, ranging from completely manual to completely automated. Basically, equipment for three separate operations is required: loading samples into plates, washing between stages and reading the end-point. Most laboratories will already possess a suitable 37°C incubator.

A variety of pipetting devices is available, of both single-shot and multi-shot types; for coating of plates with antibody or antigen the latter type is most suited because usually the requirement is for identical treatment for each well of a plate. The authors use a type that

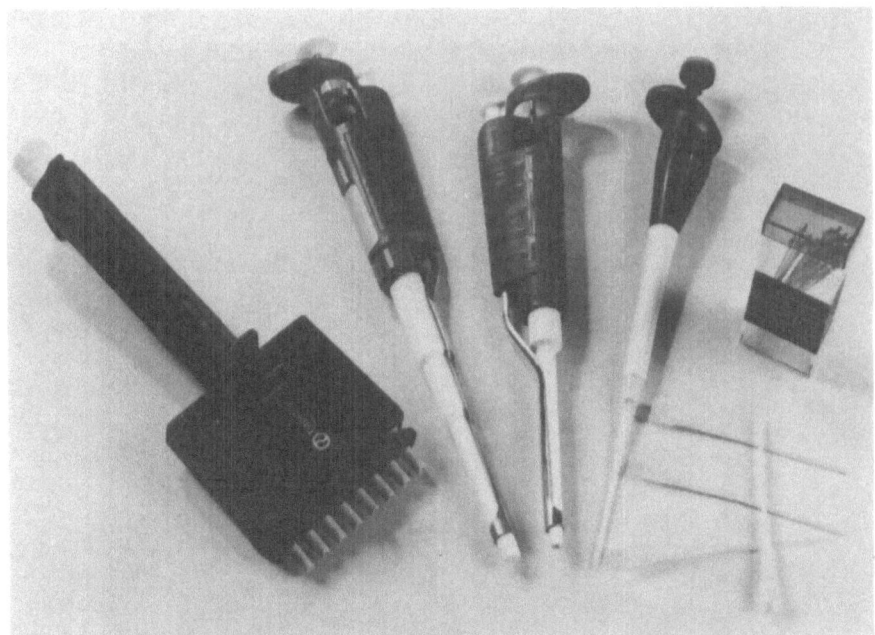

Figure 6.1
A range of single-shot and multichannel plunger-type pipettes

has a pump action and that employs disposable plastic tips (Figure 6.1). These pipettes are available from a number of suppliers in a range of sizes from $2\,\mu l$ to 5 ml and have fixed or adjustable volumes. Variations of these include types with an integral reservoir giving several deliveries before refilling and those that draw solution from an external container via a tube.

Thorough washing of plates between assay steps is crucial if consistently true results are to be achieved. This is the case particularly when enzyme-labelled antibody is used; incomplete or irregular washing at this stage can lead to spurious colour production in some wells or an unacceptably high background colour. At its simplest, washing can be carried out with a squeeze bottle containing wash buffer but it is difficult to achieve reproducible washing in this manner. A semi-automated device is available which allows alternate evacuation and filling of one column of wells at a time (Figure 6.2), using external sources of vacuum and wash buffer. The most efficient (but most expensive) equipment is an automated washer which operates on all 96 wells simultaneously. These can usually be programmed for the number of wash cycles required, with three to five cycles being the most

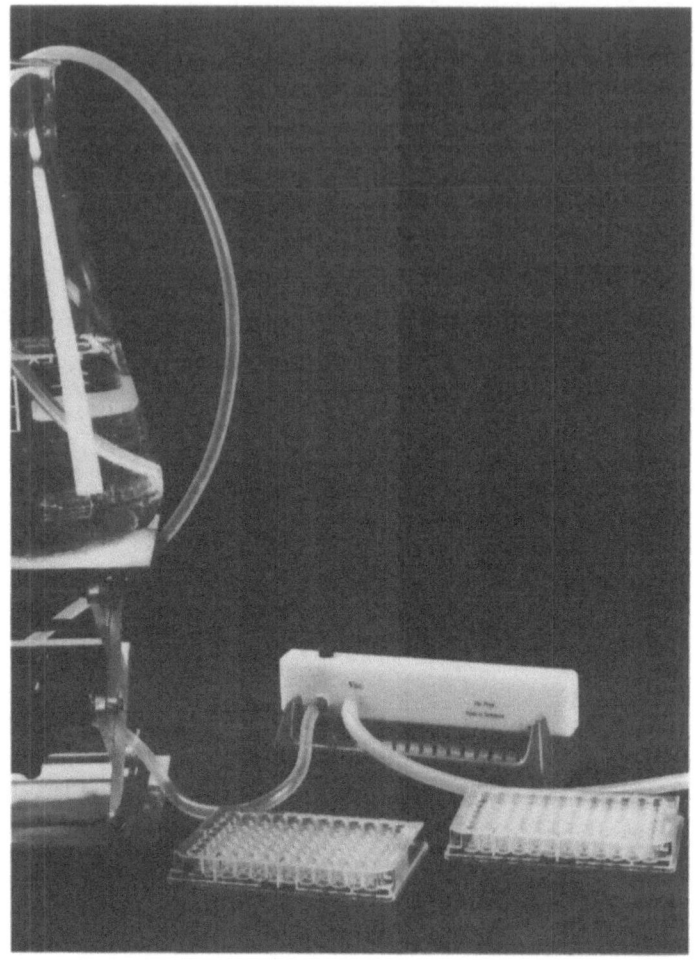

Figure 6.2
A manual multichannel plate washer with gravity-fed buffer supply

commonly used, and they can have attachments allowing a batch of plates to be processed automatically (Figure 6.3).

Reading of assay end-points can be qualitative or quantitative. In protocols employing a culture step it will not be possible to relate the end-point to the number of cells of the pathogen in the original sample, because an unknown degree of multiplication will have taken place in culture; the result has to be read as either positive or negative, so it might be thought that it is merely necessary to judge the result by eye.

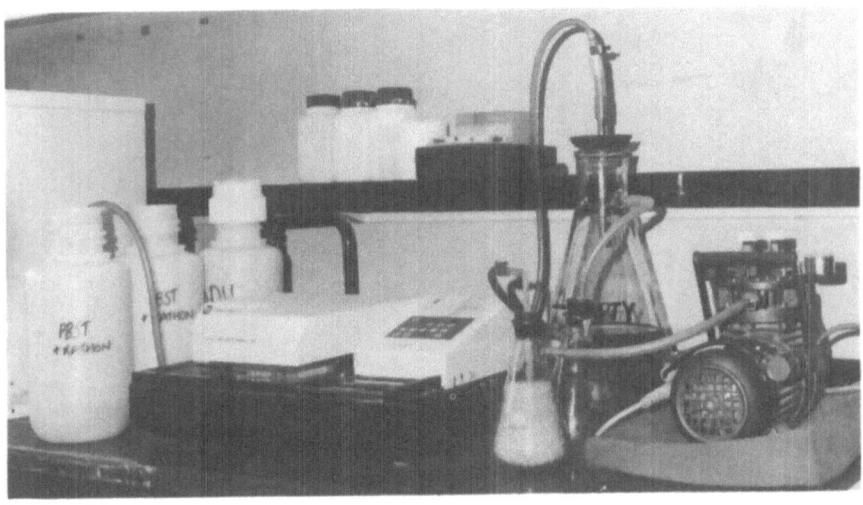

Figure 6.3
Automated plate washer with pumped buffer supply, vacuum pump and waste reservoir

However, as with any analytical technique, it is usual to include controls in immunoassays and carry out data analysis as a measure of assay performance (see Chapter 7) and for this reason quantification of end-point is preferable. Assays for microbial toxins in foods can, of course, be meaningfully quantitative and end-points need to be measured by an instrument. This is particularly true for low levels close to the assay threshold, which might not be distinguishable by eye from the background. It should be possible to transfer the contents of a microtitration plate well to a spectrophotometer cuvette for absorbance reading, but this is tedious and unsatisfactory. Dedicated microtitration plate readers are available using a selection of wavelengths that allow measurement of several different end-points (Figure 6.4); absorbance readings can be printed out or sent to a computer (see below).

All of the above operations, even with the semi-automatic equipment described, still require manual intervention. For a laboratory with a high throughput of microtitration plate assays, ELISAs in this format can be fully automated. The robotic equipment described earlier (Biomek, Beckman Instruments) for use in cell culture is designed, with appropriate attachments, to carry out all ELISA steps, including filling, washing, reading plates and data processing. With this equipment assays could, if desired, be run overnight.

Figure 6.4
Microtitration plate absorbance reader which can either be used independently or connected to a computer and printer for data analysis

---

—— 6.2 ——
DATA GATHERING, STORAGE AND PROCESSING

These aspects, as with the actual assay, can be handled in a number of ways, from manual through to computerized processes. For a simple one-off assay it is probably sufficient to store the output from a plate reader manually and to draw standard curves by hand. Results from samples can then be read off the standard curve. However, if a large number of assays are to be run, and especially if quality is to be controlled over an extended time period, then some form of computerized storage and processing is desirable.

Most laboratories will possess a desk-top computer and suitable software is available in IBM-compatible format (Titersoft, ICN Flow) to take data from a plate reader. Absorbance readings can then be printed in a graphical form to match the microtitration plate layout, and standard curves generated by the computer. Further data analysis is possible and this will be discussed in the next chapter.

A negative baseline has to be set beyond which a sample is classed as positive. The authors' usual system is to run several (>10) replicates of a blank (known negative) sample; positive results are then taken as those falling outside two standard deviations from the mean blank reading, above or below depending on the assay type (competitive or non-competitive).

CHAPTER
—7—

Assessment of Assay Performance

Whatever analytical technique is used, a good laboratory will make an assessment of the performance of that technique and, as a consequence, of the workers using the assay. This assessment should be such that both immediate and longer-term problems are seen. Some of what follows is specific to immunoassays and some is more general.

In a protocol for detection of pathogens involving both a cultural stage and an immunoassay, quality control measures should take account of both stages; for assay of toxins, any extraction procedure should also be tested.

—7.1—
SOURCES OF ERROR IN ASSAYS

Possible sources of error are varied but can be divided roughly into two groups: those causing random problems, i.e. 'one-off' errors, and those causing systematic error, leading to bias in a set of results.

First, a few examples of potential random errors are given. An obvious one is a simple mistake in pipetting, dilution, plate filling, washing or other technical procedure. This may show itself as a result blatantly out-of-step with other samples and the easiest way to overcome this is to use sufficient replication of the samples.

Problems with antibody binding are, of course, unique to

immunoassays. Unwanted cross-reactions with non-target analyte should not occur if the antibodies in the assay have been fully assessed for specificity. Non-specific binding, i.e. general 'sticking' of antibody in an assay, is usually eliminated by inclusion of a surface-active agent such as Tween-20 in the assay diluents and buffers (see Appendix 7).

Interference in binding, usually as a reduction in efficiency, by components of the food matrix can occur. High levels of protein can have non-specific interactions and it is possible for fatty acids to denature the antibody. If the assay is sufficiently sensitive, then matrix interference may often be overcome simply by diluting the sample before assay. In a system for food-poisoning bacteria in which a cultural stage is employed, then some dilution of the food into the medium will already have taken place. Additionally, growth of the organisms in the culture can lead to further breakdown of potentially interfering food components. Toxin assays usually involve an extraction step which might eliminate the interference without the need for further action. The effect of the food matrix on an assay is the most common source of problems and this aspect should be fully tested for each new type of food, as discussed below.

Some foods are known to contain the same enzymes as those used for labelling, e.g. phosphatases in meat and peroxidase in soya products. However, due to the washing procedures used in ELISAs, these endogenous enzymes should not be present at the substrate stage of the assay; problems could be encountered from this source in non-separation (homogeneous) assays.

Systematic error can be introduced into assays in several ways and, unless good quality control is practised, it may be more difficult to detect. If a reagent is ageing, there is drifting of a pipette calibration or a decline in a spectrophotometer light source, for example, then slow long-term changes in assay results can take place. Appropriate measures, as outlined below, should reveal this drift.

One problem that can occur occasionally with microtitration plate-based assays is the phenomenon of 'edge effects', manifesting itself as systematically higher or lower readings at the edges of the plate. This is usually attributed to incubating plates in stacks, whereby heat penetration to the centre is severely retarded. Plates should never be incubated in this way, two being the maximum stack for best results.

Lids should, naturally, always be fitted to prevent evaporation of contents. Placing of replicates randomly across a plate is also useful.

——— 7.2 ———
VALIDATION AND QUALITY CONTROL

Immunoassays for food-poisoning bacteria and their toxins are a relatively newly introduced concept, and will usually replace a tried and tested method. It is important therefore that the assay system is validated by comparison against the original method, and for each situation in which it is to be used.

In order to be meaningful, comparative trials of old and new methods need to be carried out across a number of laboratories and range of foods. This is probably beyond the resources of the smaller end-user, but fortunately such trials are performed periodically by various government departments or professional groups. In the USA, the usual body is the Association of Official Analytical Chemists (AOAC) and the results are published in their journal. In the UK, but to a lesser extent, collaborative trials are organized by the Ministry of Agriculture, Fisheries and Food. However, it is still a sensible idea for end-users to perform their own comparisons, albeit on a smaller scale, by running identical samples side-by-side in the two methods, until it is clear that the immunoassay system performs as well as the older method.

Within the immunoassay method, the effect of the food matrix should be assessed for each foodstuff, as mentioned above. The easiest way to do this is to prepare analyte standards in buffer, and in the food matrix/diluent at the concentration which is to be used routinely. Assay performed on both series of standards should produce identical (superimposable) curves (Figure 7.1). If an unacceptable difference occurs, then one of the measures mentioned previously can be considered.

Internal controls should be included in each batch of assays. For bacteria, standardized positive controls are difficult because, if unsuitably stored, cells in standard suspensions can multiply or die off. As a standard in work on development of a salmonella ELISA, the author has found that suspensions of pasteurized cells can be stored at $-70°C$ for lengthy periods; the assay has, of course, to recognize

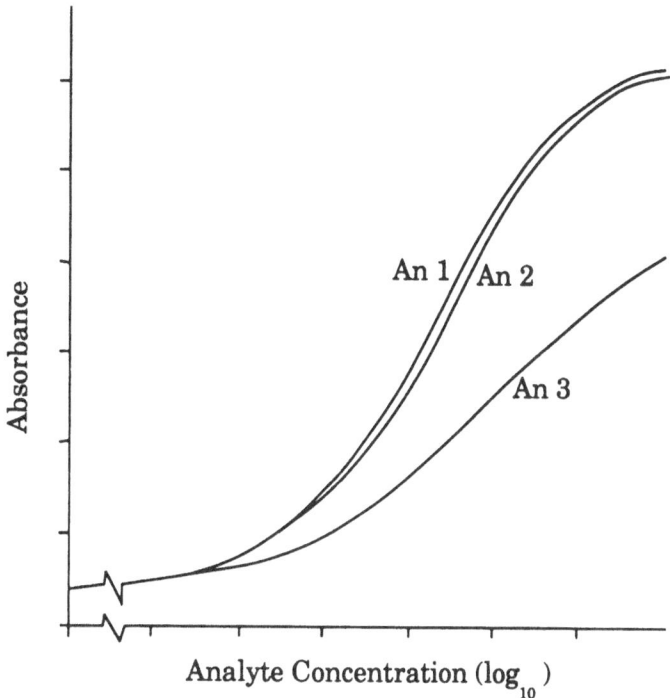

Figure 7.1
Superimposability of standard curves in an excess-reagent sandwich assay (An 1 = analyte diluted in buffer; An 2, An 3 = analyte diluted in two food extracts). The first food (An 2) shows little or no matrix effect but the second (An 3) shows considerable depression of antibody binding or inhibition of assay signal

pasteurized cells for this to be a useful method. Even these are not ideal because they do not test all parts of a system which includes a cultural stage. A few years ago, Beckers *et al.* (1985) introduced the idea of reference capsules containing a known number of freeze-dried cells. These could be added as required to a negative sample which is then passed through all stages of the assay. The capsules have the additional advantage of containing sublethally damaged organisms, thus testing recovery in the cultural stage. A known negative internal control, used for calculating the threshold value for a positive result, should also be included.

An appropriate quality control measure (to control for ageing reagents and other slow changes in assay conditions) is to use the data transformation known as a precision profile. For a full mathematical treatment of this the reader is referred to Ekins (1983), but an outline is given below. The precision profile is a graphical plot of the coefficient of variation of assay measurements plotted against concentration of analyte. From a standard curve of the assay, with each point replicated six times, the standard deviation (s.d.) of each determination of analyte concentration is calculated:

$$\text{s.d.}_{\text{analyte}} = \frac{\text{s.d.}_{\text{response}}}{\text{Slope of standard curve}}$$

The response is the absorbance or other measured signal at each analyte concentration; the slope of the standard curve is found by drawing a tangent to the curve at each analyte concentration measured. The coefficient of variation (CV) is then calculated for each analyte concentration:

$$CV = \frac{\text{s.d.}_{\text{analyte}}}{\text{Concentration of analyte}}$$

If the CV is then plotted against analyte concentration, a characteristically shaped graph is produced (Figure 7.2). Such plots should be produced at regular intervals and also when there is any change of reagents, operator or other conditions. Comparison of the current graph with accumulated historical graphs can show any drift in the assay. For microtitration plate-based assays, within-plate and between-plate comparisons should be made by this method for each batch of plates used.

A further use for this technique is to define a working range for the assay. If a threshold value of CV (often 10%) is regarded as acceptable, then the graph will show within which span of analyte concentrations a result must fall in order to achieve this precision. This latter aspect is less critical for the assay of bacteria rather than for that of toxins, due to the unknown degree of multiplication in the cultural stage.

Calculation of precision profiles is possible by hand, but tedious, and suitable computer software is available (Immunofit, Beckman Instruments) which will perform the calculations and produce the profiles far more easily.

—— 7.3 ——
RELATIVE COSTS OF IMMUNOASSAYS

When considering the costs of any assay system, several factors need to be taken into account. The direct costs, such as reagents, equipment

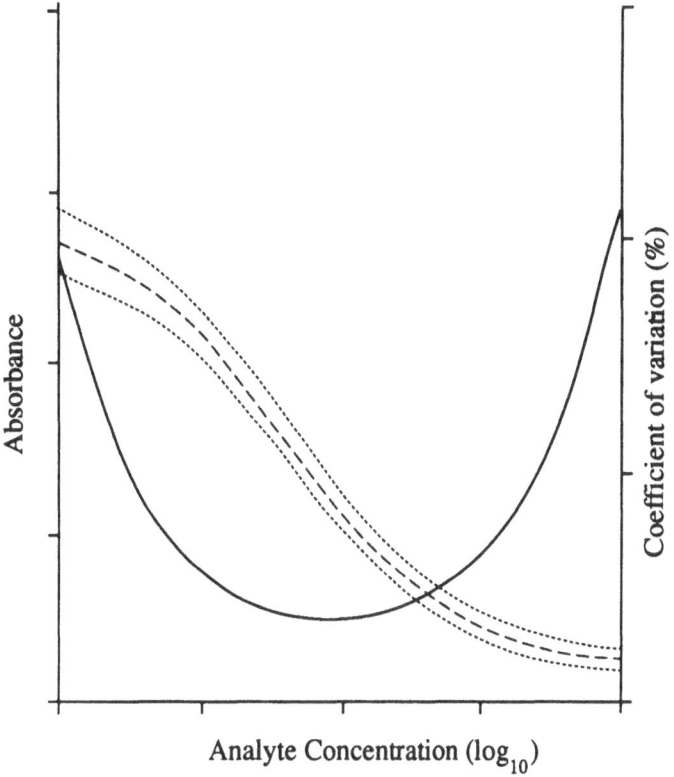

Figure 7.2
An idealized precision profile. The graph shows a standard curve (dashed line) for a competitive assay with confidence limits (dotted lines), both plotted against absorbance; the precision profile (solid line) is a plot of the coefficient of variation for the standard curve

and labour, are obvious considerations but there are also a number of indirect costs, which may initially be less apparent. These would include factors such as the cost of delay while the result of an analysis is obtained, and the consequences of a failure in the reliability of an assay. The former factor may require expensive cold storage for some food products; the latter could entail huge costs if a product has to be recalled owing to a false negative result, these costs being both monetary and the less quantifiable loss of public confidence in the company's products.

As discussed in Sections 3.2 and 3.3, traditional methods for detection of bacteria and their toxins are lengthy and labour intensive.

Inevitably, then, the costs are high, and immunoassay systems offer considerable scope for reduction in those costs. This can be illustrated by taking detection of salmonellae as an example. The full cultural method (see Figure 5.21) takes at least 4 days and thus one of the indirect costs mentioned above, product holding-time, will be high if there is to be certainty that the product is salmonellae free before release. A multiplicity of culture media is used which, in addition to the cost of ingredients, is labour-intensive to produce. Again, the several transfers required between media are laborious and thus expensive. An alternative, antibody-based protocol (see Figure 5.22) uses only one cultural step and is completed in a much shorter time, offering considerable savings on both the direct and indirect costs of the traditional method.

Equipment costs will vary according to assay format and the types of equipment that might be needed have been covered.

Although the research costs of producing a specific antibody are high, a monoclonal antibody, once developed, can be produced in almost unlimited amounts. The initial cost is quickly diluted in use, and long-term reagent costs for immunassays are thus quite low, enzymes and substrates being no more expensive than most other assay materials. Many analytical laboratories will not be able to produce their own antibodies, so wider use of immunoassays for microbial detection may depend on reasonably priced reagents being commercially available.

—— 7.4 ——
LIMITATIONS OF IMMUNOASSAYS

A number of potential limitations of immunoassays have been touched on already. For example, antibody production is a specialized procedure and those antibodies developed must have the desired specificity or the assay will not be acceptable. However, when a finished assay is produced the probable advantages will usually outweigh the limitations heavily.

However, there is one rather more philosophical point that is of concern. Food analysts have been rather slow to take up immunoassays for routine use and there are a number of reasons for this. Many analysts are chemists by training and seem to find it hard to accept a biologically based assay, perhaps through a lack of understanding of the basis of the technique. Also, as mentioned previously, conventional methodology is often considerably less than perfect. This means that all 'true positives' have not been detected in the past. A new method which displays greater sensitivity is likely to lead to a higher rejection rate in food production, which could be economically unacceptable to a company. It is possible that for this reason the conventional methodology will be difficult to displace. It is unlikely, however, that this would be admitted publicly!

BIBLIOGRAPHY

Immunoassays for Veterinary and Food Analysis (1988) Eds B. A. Morris, M. N. Clifford and R. Jackman, Elsevier Applied Science, London, UK

CHAPTER
— 8 —

Assays Available Commercially

Commercial assay production is a fast-developing field and thus any discussion will most probably be very quickly outdated. For this reason a few examples, based on different principles, have been chosen to illustrate the kind of assay available. Others are marketed and the authors would like to point out that they are not necessarily recommending the assays discussed to the exclusion of the alternatives.

—— 8.1 ——
ELISAs

Three examples are considered here, each with colorimetric end-point detection. Two are for bacterial cells and the third is for enterotoxin.

The first is the TECRA Salmonella Visual Immunoassay (Bioenterprises Pty Ltd, Roseville, NSW, Australia) which has received interim approval from the AOAC (Flowers *et al.*, 1988) as a screening method for detection of salmonellae in all food types. The assay is an excess-reagent sandwich type (see Section 5.8.1), based completely on polyclonal antibodies, and the result is determined visually. The antibodies are raised to flagellar antigens and thus will not recognize the very few non-flagellated salmonella serotypes that exist, such as *Sal. pullorum*. A modification of the standard enrichment protocol is used, employing 18–22 h pre-enrichment culture, either 6–8 h or 16–20 h

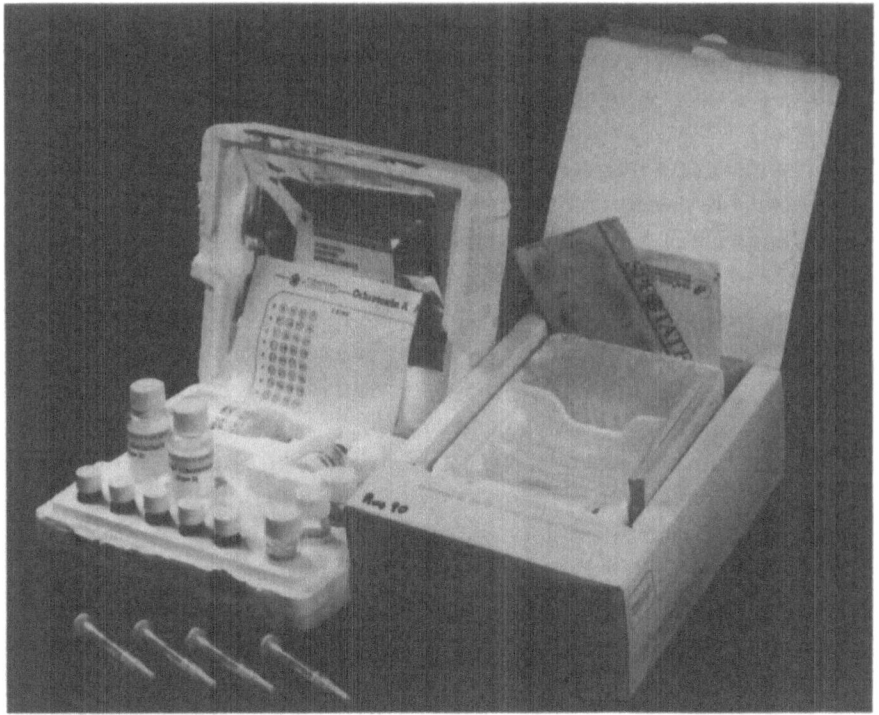

Figure 8.1
Presentation of commercially available kits based on antibody technology. In the foreground are
some immunoaffinity columns

selective enrichment culture and finally either 16–20 h or 6–8 h post-enrichment culture. The alternative times for the second and third cultures are for foods normally expected to carry low and high microbial loads respectively.

All materials for the ELISA are supplied packaged in a box (Figure 8.1). The assay is based on a modified 96–well microtitration plate format. The wells are supplied as strips of 12 which can be inserted as desired into a base plate; this arrangement enables a small number of tests to be carried out while still allowing the use of equipment, such as washers, based on the 96–well plate format. The wells are pre-coated with the capture antibodies. Positive (purified salmonella antigen) and negative (lactose) controls, horseradish peroxidase-conjugated antibody and substrate (ABTS) are each supplied lyophilized in vials, together with appropriate diluents. A wash solution concentrate is supplied, to be diluted in water to working strength. To perform the assay, aliquots of post-enrichment broth are heated in boiling water for 15 min before transfer to the wells. After two 30 min incubation steps, with appropriate washes, substrate is added and the assay is finished by the addition of a stop solution (aqueous NaF). The end-point colour will be a shade of blue-green. Sample and control wells are read by comparison to a colour chart supplied with the kit.

The second ELISA to be described is the Listeria-Tek assay (Organon Teknika Ltd, Cambridge, UK) which has an instrumentally determined end-point. The assay is entirely monoclonal antibody based and is a sandwich ELISA with the two antibodies recognizing different epitopes. It is specific to the genus *Listeria*, detecting *L. monocytogenes* and several other *Listeria* species. Conventional listeria enrichment protocols are followed, using antibiotic-supplemented media appropriate to the food sample, incubating for up to 48 h.

Presentation of the kit is similar to the TECRA Salmonella assay, with lyophilized reagents in vials together with diluents, wash concentrate and stop solution (H_2SO_4). The antibody coated wells are in 8 strips of 12, with a strip holder. The antibody label is again peroxidase but the substrate is a TMB/hydrogen peroxide system. Aliquots of the final broth culture are heated in a boiling water bath for 20 minutes and cooled before addition to the wells. Enzyme-labelled antibody is added, sample and label being incubated together in a single 1 h step. After washing and substrate addition, a blue colour is produced which is turned yellow by the addition of the stop solution. The colour is read on a microtitration plate reader at 450 nm. The analyst is directed to calculate a cut-off value beyond which results are considered positive; this value is the mean absorbance of the negative control + 0.15 absorbance unit.

The third test kit is a competitive ELISA (see Section 5.8.1) for *E. coli* heat-stable enterotoxin, available from Oxoid (Unipath Ltd, Basingstoke, UK). Microtitration plate wells are pre-coated with synthetic toxin produced by peptide synthesis and a single horseradish peroxidase-conjugated monoclonal antibody is used. The wells are packaged as 6 × 16 format, held in a frame. The usual controls, diluents, wash buffer and stop solution (H_2SO_4) are supplied. The substrate is an *o*–phenylenediamine/hydrogen peroxide system. The assay is performed on culture filtrates from suspect *E. coli* strains isolated from clinical samples. The sample and enzyme-conjugated antibody are incubated together in the wells for 90 min. After washing, substrate and stop solution are added in the usual way. The manufacturers recommend that clear positive or negative results be read by eye, in comparison to the controls. Because this is a competitive format, a yellow colour indicates a negative and a colourless well indicates a positive. Ambiguous wells should be read by a plate reader at 490 nm. The absorbance is adjusted based on the values given by the controls using a supplied equation, and a cut-off value of 0.2 unit then applied.

—— 8.2 ——
OTHER FORMATS

Two kits are described below based on different uses of antibodies, one for a bacterial toxin and one for cells.

The first is a reversed passive latex agglutination (RPLA – see Section 5.8.2) system for *Bacillus cereus* diarrhoeal-type enterotoxin in foods, available from Oxoid (Unipath Ltd, Basingstoke, UK). The kit contains latex sensitized (i.e. coated) with rabbit polyclonal antibody to the toxin, plus controls and diluent. A simple procedure is used to extract the toxin from food: 10 g is blended in saline, centrifuged and filtered to remove remaining particulate material. Volumes of filtrate are added to V-shaped wells and a dilution series performed. After addition of the sensitized latex, the wells are agitated and left for 20–24 h at room temperature before examination for agglutination in the bottom of the V-shape. This is compared to an illustration provided in the kit. The sensitivity of the test is 2 ng/ml test extract but concentration of the food extract could give better sensitivity.

The final kit to be described is the Bio-Control 1-2 (TEST) (Bio-Control Systems Inc., Bothell, Washington, USA) which has gained initial approval by the AOAC (Flowers and Klatt, 1989) for detection of motile salmonellae in foods. The test is an immunodiffusion system carried out in a disposable plastic unit that has two chambers. The vertical chamber contains a semi-solid motility medium and is connected to a side-arm containing a tetrathionate-based selective enrichment broth. The principle is that after inoculation motile cells move from this broth into the motility gel and meet polyvalent anti-salmonella flagellar antibodies diffusing from the top; a visible band of immobilized cells forms in the meeting zone.

After a conventional pre-enrichment step, this culture is inoculated into the side-arm. The antibody solution provided is added to the top of the vertical chamber. The unit is incubated at 35°C for a minimum of 8 h and then examined in a strong light. The presence of a U-shaped or meniscus-shaped band in the upper half of the vertical chamber indicates a positive result.

CHAPTER
— 9 —

Conclusions

CRITICAL COMPARISON WITH OTHER METHODS

A comparison of antibody-based methods with other rapid methods divides into two, with different viewpoints for cells and toxins. For toxins there really are no rapid methods, other than antibody systems, that can improve on older methods of toxin detection (see Section 3.3 and Lee and Morgan, 1990). Indeed, cases can be envisaged in which antibody-based methods are the only practicable ones available for specific toxin detection. However, for cells a number of alternative technologies have emerged.

A degree of time reduction for salmonellae detection has been achieved with a repackaging of conventional cultural methods (Salmonella Rapid Test; Oxoid). This employs a plastic unit rather similar to the 1-2 TEST described earlier. A system of compartments allows the inclusion of an elective medium, two selective enrichment media and two indicator media in the same unit. Two days are required for conventional pre-enrichment culture and use of the test unit. Presumptive positives suggested by reactions in the indicator media are tested further using a latex test from the same company. This test system, although neatly presented, unfortunately suffers from the same disadvantages (see Sections 3.2 and 5.0) as the original cultural methods from which it is derived. Also, although labour-saving compared with 'in-house' media preparation, this is partially offset by the high cost of purchasing the system units.

There are various rapid methods for total numbers of bacteria which are well developed, based on electrical changes during culture (Firstenberg-Eden, 1986), enzymic detection of microbial adenosine triphosphate (ATP; Stanley, 1989), and direct fluorescent staining (Pettipher and Rodrigues, 1982). Some progress has been made with the first mentioned towards specific pathogen detection by using selective media in the system but, again, because the media are not completely selective the usual problem of non-specificity arises; mention was made earlier (see Section 5.8.5) of the introduction of antibodies into that system to improve specificity. Detection of microbial ATP, while rapid, is completely non-specific and not relevant to the present situation unless it could be combined with antibody capture.

An interesting development in the field of rapid methods is the use of a bacteriophage to introduce a label or marker into the genome of a bacterium. A group of organisms known as photobacteria occurs naturally – these are luminescent during growth. The genes responsible for this phenomenon, designated *lux* genes, have been isolated and introduced into the genome of bacteriophages. The phages themselves do not possess the full pathways for light production but, upon infection of bacteria, expression in the host produces bioluminescent cells. The luminescence can be measured in a suitable instrument, with a sensitivity of $10^2–10^3$ organisms/ml (Stewart, 1990). In order to achieve the required level of detection for pathogens in food, a cultural stage would be necessary prior to application of the phage test. The principal problem with this system is the host range of the phage vector. Many phages have a very narrow specificity, with a range that is subspecies or even subbiotype. This is far too narrow to achieve coverage of all strains of a target pathogen and, to be useful, phages of the correct specificity will need to be isolated. Creation of phages of a desired specificity by genetic manipulation may, of course, be possible in the future but as yet this is a considerable way off. Other problems of this technology include maintenance of phage stocks, and ensuring purity is not compromised by mutation of the phages. Additionally, the process technically involves the creation of recombinant bacteria and thus falls under the legislation designed to control genetic manipulation in many countries.

At present, the most likely technology to rival antibody-based methods for detection of pathogenic cells is that based on nucleic acid

92

probes. The principle involved is that extracted DNA or RNA sequences specific to a bacterial species are labelled with a marker carried by a complementary sequence of nucleic acid. Using a sequence with a high number of copies per cell, such as messenger RNA, can give a considerable signal but, again, to reach the target detection limit in foods a cultural stage is necessary. An alternative system under development is to amplify the extracted nucleic acid by a method such as the polymerase chain reaction (PCR; Oste, 1988), rather than replicating the living cells. However, PCR is a technique that is rather sensitive to interference and not suited to routine laboratory analysis.

The amount of research work involved in development of a nucleic acid probe is greater than that required for monoclonal antibody production, and the rather involved extraction and hybridization techniques used in application of probes do not readily lend themselves to automation. Nevertheless, the potential specificity of targeting the very material that is responsible for the structure and function of an organism is attractive and much work continues in this area.

Currently, then, systems based on antibody technology lead the field in specific pathogen detection. The unique ability of antibodies to bind physically to specific bacterial cells can be exploited in many ways. Interfering substances can be removed from the vicinity of the bound bacteria and the cells can, if necessary, actually be transferred to another location. This specific physical binding is the major advantage of antibody usage.

—— 9.2 ——
FUTURE IMPACT OF NEW TECHNOLOGY

9.2.1 Impact on public health

In Chapter 1 the authors touched on the idea that the labour-intensive, and thus expensive, traditional microbiological methods for detection of pathogens were a considerable disincentive to wider testing of foods. One of the recommendations of the Richmond Committee in the UK (HMSO, 1990, 1991) was that an extensive monitoring and surveillance system be set up to combat the apparently steeply rising incidence of food poisoning. If this proposed system is to have any impact on public health, then the methodology used must be cost-effective, otherwise the surveillance is unlikely to be sufficiently wide to be of use. Rapid microbiological methods offer a means of achieving this cost-effectiveness. Additionally, new responsibility was laid down by the resultant Food Safety Act that producers ensure food is not 'injurious to health'; the wider testing of materials and products encouraged by new technology should lead to safer foods.

9.2.2 Impact on industrial practice

In a thought-provoking book entitled *Food Microbiology – A Framework for the Future*, published more than a decade ago (Sharpe, 1980), Anthony Sharpe called for an alternative approach to food microbiology. His main concept is that potential deterioration of food subsequent to production is predictable and should be the main parameter determined. However, a subsidiary theme of the book is a plea for increased instrumentation and automated analyses in microbiology. He suggests that microbial parameters should be measurable on-line in the same way as chemical or physical quantities such as pH. Although this day has, even now, not arrived, some of the systems discussed in the present book are moving towards this idea.

An intermediate measure, between end-product testing and true on-line monitoring, is the Hazard Analysis Critical Control Point (HACCP) Scheme. This is already in operation in some food production systems, and wider use in the UK was another of the recommendations of the Richmond Committee. After an analysis of production hazards, critical points in the system are identified and subjected to careful monitoring on a routine basis. With a food known to be frequently involved in bacterial poisoning, such as poultry, these points may be numerous and thus the number of tests considerable. Again, more convenient and less expensive technology should make a considerable impact on the speed of introduction of the HACCP scheme.

Throughout this book the specificity and versatility of antibody-based systems have been emphasized. Technology derived from the clinical area has been successfully transferred to food analyses and it is to be hoped that its use will extend further.

BIBLIOGRAPHY

Foodborne Microorganisms and their Toxins: Developing Methodology (1986) Eds M. D. Pierson and N. J. Stern, Marcel Dekker Inc., New York, USA

Gene Probes for Bacteria (1990) Eds A. J. L. Macario and E. C. de Macario, Academic Press, San Diego, USA

Appendices

Most of the methods described in this section have been used in the author's laboratory. If modifications have been made to the original method then this will be made clear.

There will, of course, be alternative methods available in some instances but those described have mostly been found to be useful and successful.

Appendix 1:
DEFINITION OF TERMS USED

affinity   the energy of binding of an *antibody* and *epitope*

agglutination   formation of a visible lattice of bacterial cells or
                        antibody-coated particles as a consequence of the
                        binding of antibody to antigen

analyte   the target compound/molecule/cell to be detected
                        or measured

antibody   a protein molecule (immunoglobulin) which binds
                        specifically to another molecule (see *epitope*, *antigen*)

antigen   the structure to which antibodies bind; can consist
                        of many *epitopes* each of which is the target for a
                        specific antibody

anti-species antibody	in the context of immunoassays, an antibody with a *label* attached which binds to, and signals the presence of, all antibodies of a particular species
carrier protein	a protein to which a normally non-immunogenic molecule is coupled to elicit an immune response (see *hapten*)
cross-reactivity	the degree of binding of an antibody to a non-target molecule
displacement	the lowered binding of antibody to antigen in a *limited-reagent immunoassay* caused by the presence of analyte in the sample
EIA	enzyme immunoassay
ELISA	enzyme-linked immunosorbent assay
end-point	a signal from an assay which is used to quantify the *analyte*
epitope	the specific site on a molecule to which the antibody binds; can be as small as three to six amino acid residues
excess-reagent immunoassay	an assay which uses an excess of the antibody reagents, usually in the form of a 'sandwich' in which the analyte is sandwiched between two different antibodies. Only possible with analytes that have more than one *epitope*
Fab fragments	the region of antibody structure responsible for specific binding to the *epitope*
Fc fragments	the region of antibody structure responsible for effector function
flow cytometer	an instrument which determines the number and properties of cells in a flowing stream
hapten	a molecule which is normally non-immunogenic but which becomes so when linked to a *carrier protein*
heterogeneous immunoassay	an assay in which component parts are physically separated by being either immobilized on a solid phase or in solution
homogeneous immunoassay	an assay in which all component parts are in the liquid phase
idiotype	the region of antibody structure which is chemically unique to that antibody

immunoaffinity column	consists of an antibody-coated gel or similar support matrix packed into a glass or plastic column
immunogen	a preparation which will be recognized as foreign by an animal immune system, stimulating the production of antibodies
immunosensor	a device producing an electrical or optical signal when antibody binding occurs
label	a tracer molecule attached to an antibody or analyte
isotype	the class of antibody structure, e.g. IgG, IgM
ligand	in this context, a molecule which binds specifically to another molecule, e.g. antigen–antibody binding
limited-reagent immunoassay	an assay in which one of the components is present in a limiting concentration thus leading to competition for antibody binding sites. Also called a competitive assay
monoclonal antibody	a preparation containing antibodies of one chemical type, derived originally from a single line of lymphocyte cells. Monoclonal antibodies are of a single *specificity* and *affinity*
matrix	the mix of components forming a food
polyclonal antibody	a preparation, usually an animal serum, containing a mixture of antibodies of different specificities and affinities
phase separation	techniques for separating free antibody from that bound to target analyte
precipitation	production of insoluble complexes by binding of antibody to soluble antigen
selective enrichment culture	growth of bacteria in a medium designed to support one species whilst suppressing the growth of others
sensitivity	the limit of detection of an analyte in an assay
specificity	the range of targets to which a given antibody will bind; may be broad or narrow

Appendix 2:
HAPTEN CONJUGATION METHODS

Several proteins are suitable as carrier molecules but in the authors' laboratory bovine serum albumin (BSA) or keyhole limpet haemocyanin (KLH) is usually used. The site of conjugation is important to the specificity of the antibodies produced because structures furthest from the coupling groups will be most strongly recognized. It will usually be necessary to apply standard procedures to remove unconjugated materials after conjugate preparation.

A2.1 Conjugation of peptides

A number of methods are available in which a carboxylic group on the haptenic molecule is coupled to the protein via amide bond formation. Examples of those that have been used in the authors' laboratory (Wilkinson *et al.*, 1991) are given below. The method of choice will depend to some extent on the solubility of the peptide.

A2.1.1 *Mixed anhydride reaction*

The compound to be conjugated (25 mg) is dissolved in re-distilled dry dioxane (2 ml) and cooled to 12°C. Tri-*n*-butylamine (0.05 ml) is added and the solution is left for 20 min at the same temperature; *iso*-butylchlorocarbonate (0.5 ml; again re-distilled and dry) is added and the solution stood for a further 20 min at 12°C. The carrier protein is dissolved (7–10 mg ml^{-1}) in water:dioxane (2:1, v/v) and added (7 ml) to the above solution. The mixture is rapidly adjusted to pH 8.0 with 0.1 M sodium hydroxide and stood for 4 h at 4°C. The product is extensively dialysed against water before lyophilization or other storage procedure.

A2.1.2 *Carbodiimide reaction*

The haptenic molecule (5 mg) is dissolved in phosphate buffer pH 7.0 (0.1 M, 3 ml). The carrier protein is dissolved (10 mg ml^{-1}) in 0.1 M sodium chloride solution and added (5 ml) to the hapten solution together with 1-ethyl-3-(dimethylpropyl)-carbodiimide (25 mg). The mixture is stirred for 24 h in the dark at 20°C before clean-up procedures are applied.

A2.2 Conjugation of sugars

A number of methods are available for coupling carbohydrates to protein. Complex procedures involving formation of *p*-aminophenyl derivatives followed by coupling using diazonium or phenylisothiocyanate reactions have been published (McBroom *et al.*, 1972). It is simpler to use the periodate oxidation method (see Appendix 6, A6.1.1; see also Butler and Chen, 1967).

A2.3 Conjugation of other molecules

In the present context, it is unlikely that anything other than molecules with peptide or carbohydrate residues will be used as immunogens. However, if necessary, it may be possible to produce a carboxylic acid group by modification of other structures on the molecule. Coupling via different procedures is also possible. It would be necessary to consider carefully the structure of the molecule in question. For a review of coupling procedures see Wilkinson *et al.* (1991).

Appendix 3:
PREPARATION OF ANTIGENIC STRUCTURES FROM BACTERIAL CELLS

The amount of an antigenic structure present on the cell will vary with factors such as strain of the organism, culture medium components and growth temperature. For maximum production these factors will need to be optimized for each organism used.

A3.1 Extraction of lipopolysaccharide (LPS) with trichloroacetic acid

Dried cells (10 g) are suspended in cold water (500 ml; 4°C) and mixed with 500 ml 1 M trichloroacetic acid before stirring for 3 h at 4°C. Cell debris is removed by centrifugation at 10 000 × *g* for 30 min. The supernatant is dialysed against running tap water overnight and concentrated to less than 50 ml using a rotary evaporator or similar equipment. Ethanol precipitation is then carried out, with the required LPS complex precipitating at between 50% and 85% (v/v) ethanol (Sutherland and Wilkinson, 1971).

A3.2   Phenol–water extraction of LPS

Dried cells (10 g) are suspended in water (175 ml) at 65°C. An aqueous solution of phenol (90%, w/v; 175 ml) is added and stirred vigorously for 5 min. The mixture is cooled rapidly and phase separation achieved by centrifugation at 3000 × g for 30 min at 0°C. The aqueous (upper) layer is removed and extensively dialysed (48 h in running tap water) to remove remaining phenol. After centrifugation at 100 000 × g the LPS is deposited as a clear viscous gel above the pellet at the bottom of the tube. Repeated removal, solubilization in water and ultracentrifugation give a purified product (Sutherland and Wilkinson, 1971).

A3.3   EDTA extraction of LPS

Dried cells (5 g) are suspended in 100 ml solution containing NaCl (0.15 M) and EDTA (0.05 M). After gentle rotary shaking for 10 min, the cells are pelleted by centrifugation. After extensive dialysis in running tap water, ethanol (30 ml) is added and the precipitate collected by centrifugation and dissolved in distilled water.

A3.4   Mechanical removal of flagella

Cells are harvested from the growth medium and sedimented by centrifugation. The cells are then resuspended in water (approx. 50 g wet pellet/1) and shaken vigorously for 10 min; this can be done with a mechanical mixer or blending device. Intact cells are removed by centrifugation at 5000 × g for 30 min, and debris at 16 000 × g for 20 min. Flagellar filaments are then obtained by centrifugation at 40 000 × g for 3 h (Smith and Koffler, 1971).

A3.5   Acid extraction of flagellar protein

Cells are harvested from the growth medium by centrifugation and a thick suspension prepared in saline (8.5 g/l NaCl). Flagellar protein is solubilized by adjusting the suspension to pH 2.0 with HCl (1 M) and stirring at room temperature for 30 min. Cells are removed by centrifugation at 5000 × g, and insoluble material at 100 000 × g for 1 h. The pH is then adjusted back to 7.2 with NaOH (1 M) and the solution brought to 65% saturation with $(NH_4)_2SO_4$ with the usual slow addition and stirring. After holding overnight at 4°C, the precipitated flagellin is removed by centrifugation at 15 000 × g for 15 min, dissolved in distilled water (5 ml) and extensively dialysed (Ibrahim *et al.*, 1985).

A3.6 Detergent extraction of outer membrane protein (OMP)

Cells are harvested from the growth medium by centrifugation and washed in buffer at 4°C (0.05 M Tris/HCl, pH 7.5). Cells are then resuspended in the buffer and sonicated for four 30 s periods, allowing intermediate cooling times. Debris is removed by centrifugation at 5000 × g and membranes obtained by further centrifugation of the supernatant at 100 000 × g for 1 h. The pellet formed is suspended in detergent solution (N-lauroylsarcosine sodium salt 10 g l^{-1}; EDTA, 7 mM, pH 7.6) at a protein:detergent ratio of approximately 1:4 (w/w) for 20 min. After centrifugation at 100 000 × g for 2 h, the pellet is again extracted in detergent solution and the final pellet washed in the Tris buffer (Newell *et al.*, 1984).

A3.7 Extraction of teichoic acids

Cells are harvested from the growth medium and a preparation of cell walls made using a French pass or other disintegrator, followed by appropriate centrifugation steps. Cell walls (10 g) are suspended in water (250 ml) at 4°C and 25 g trichloracetic acid gradually stirred in, stirring continuing for 48 h. Debris is removed by centrifugation at 100 000 × g for 30 min, washed with 50 ml trichloracetic acid solution (100 g l^{-1}) and the supernatants combined. Acetone (235 ml) at 0°C is added and the preparation held at that temperature for 24 h. The precipitate formed is recovered by centrifugation at 10 000 × g for 30 min, washed with acetone and ether and then freeze-dried. The lyophilized material is dissolved in 3 ml water, centrifuged at 2000 × g for 15 min and the supernatant, which contains the acids, freeze-dried for storage (Sutherland and Wilkinson, 1971).

Appendix 4:
MONOCLONAL ANTIBODY TECHNIQUES

There are currently many slight variations on the basic methods but those given below are ones that the authors have found successful.

A4.1 Production of hybridomas

For best results 10^6–10^7 myeloma cells and 10^7–10^8 spleen (lymphocyte) cells are needed. A myeloma cell line (the authors use

X63.Ag8.653, Flow Laboratories) is grown in flasks of culture medium (Optimem; see A4.5) at 37°C in an incubator with an atmosphere enriched with 5% (v/v) carbon dioxide. A suitably immunized mouse (see Section 5.1) is given a booster injection 4 days prior to the fusion. Spleen cells are washed out in Optimem using a syringe and needle to flush the tissue through repeatedly, made up to about 50 ml with Optimem and centrifuged in a bench-top machine at 1500 rev./min. The myeloma cells are also harvested by centrifugation and the numbers of each cell type determined by resuspending in 10 ml Optimem and using a microscope counting chamber. The two populations are mixed in a ratio of myeloma cells:spleen cells between 1:1 and 1:10, then centrifuged at 1500 rev./min for 5 min. The supernatant is discarded and the tube manipulated gently to loosen the pellet. Filter-sterile polyethylene glycol solution (grade 1500, 400 ml l^{-1}, 0.8 ml) is added slowly over 1 min with very gentle swirling of the mixture. After standing for 1 min, 1 ml Optimem is added over a further minute *without* mixing; further slow dilution with 20 ml Optimem is carried out over the next 5 minutes. The fused cell mixture is harvested by centrifugation at 1500 rev./min for 15 min. The pellet is resuspended in 50 ml Optimem containing double-strength HT (hypoxanthine, thymidine) solution (40 ml l^{-1}; Flow Laboratories, Irvine, Scotland, UK). The suspension is then distributed into the inner 60 wells (200 μl/well of a 6 × 10 array) of four sterile microtitration plates; this can be done manually but the authors use the Biomek robotic equipment (see Section 5.3). The plates, with lids, are placed in the carbon dioxide incubator for about 3 h.

After this period, 100 μl medium from each well is replaced with fresh culture medium, this time containing double-strength HAT (hypoxanthine, aminopterin, thymidine) solution (Flow Laboratories) instead of HT. The plates are reincubated and the next day the medium is again changed, using normal-strength HAT solution (20 ml l^{-1}) in the culture medium. Medium changes are repeated on alternate days and feeder cells (red blood cells, 1:1000, v/v) added to the medium on about day 4.

A4.2 Screening for antibody production

After about 14 days' incubation, all unfused spleen and myeloma cells should have died, and clumps of hybridoma cells should be visible with a low-power microscope. To screen for antibody production 100 or

150 μl volumes of culture medium are removed to an antigen-coated microtitration plate (see Appendix 7); this should only be done when 5 days or more have elapsed since the last medium change, allowing time for antibody concentration in the medium to increase. Cells from wells positive for antibody should be expanded by transferring into the larger wells of 48-well or 24-well plates; small increases in volume are preferable to larger steps.

A4.3 Cloning

At this stage it is quite probable that the cell cultures will be polyclonal; thus it is necessary to begin the process of isolating monoclonal lines. This is done by a limiting-dilution procedure in which cell suspensions are diluted in microtitration plates to the point at which the highest dilution is statistically likely to contain only a single cell in each well. Again, this can be done manually or by the robot. The process starts with 100 μl of a suspension containing about 5000 cells ml^{-1} placed in each well of the first column of a 96-well plate. The suspension is then serially diluted in Optimem in 100 μl + 100 μl steps along the plate, giving one column of replicates at each dilution level; for a 10 × 6 well array the series therefore runs from 500 cells/well in the first column to about 1/well in column 10. The clone plates are again incubated, with medium changes as previously. At this stage HAT can be omitted from the medium but it is advisable to give at least one change with a medium containing HT, to allow the aminopterin to be diluted out. Screening and cloning are repeated as necessary. For production of larger quantities of antibody, various sizes of flasks are available; volume increase should again be in small step sizes.

A4.4 Preservation of cell lines

At each stage, for reasons discussed in Section 5.3, samples of cell lines should be preserved by freezing. The medium used is fetal calf serum containing dimethylsulphoxide (DMSO, 40 ml l^{-1}) as a cryoprotecting agent. Cells are harvested by centrifugation and resuspended in cold (0°C) medium before pipetting into plastic stoppered cryotubes (approximately 10^6 cells in 0.25 ml medium is suitable). The tubes are frozen overnight at −70°C then transferred to a liquid nitrogen storage system. Cells should always be in a healthy state for freezing, and exposure to DMSO at room temperature must be kept to a minimum.

To revive cells, the tubes are thawed quickly in a 37°C water bath and cells diluted with cold Optimem (20 ml). After centrifugation at 1500 rev./min for 5 min, the supernatant is discarded and the cells transferred to a suitable culture plate with fresh medium and incubated.

A4.5 Culture medium

A number of different media are available for hybridoma cell culture. The authors use Optimem I from Gibco Ltd (Renfrew, Scotland, UK); where Optimem is mentioned in the preceding sections, the following composition is used:

Optimem I	1 tube (as supplied)
Double glass-distilled water	1 litre
Mercaptoethanol (50 mM solution)	1 ml
$NaHCO_3$	2.4 g
Penicillin/streptomycin solution (see below)	40 ml

Adjust pH to 7.2–7.4 and filter sterilize
Before use add fetal calf serum (40 ml l^{-1}; Gibco Ltd)

Penicillin/streptomycin solution:	
Penicillin (Sigma Chemical Co.)	5000 units ml^{-1}
Streptomycin (Sigma Chemical Co.)	5 mg ml^{-1}
dissolved in glass-distilled water and filter sterilized.	

All media and medium components are held either frozen ($-20°C$) or refrigerated, depending on the length of storage required. As a further precaution against contamination, it is advisable to filter sterilize the medium again immediately before use in cell culture.

Appendix 5:
ENZYME/SUBSTRATE SYSTEMS

The systems given below are those that the authors have found most useful.

A5.1   Colorimetric substrates

A5.1.1   Horseradish peroxidase (HRP) substrates

1. 3,3',5,5'-Tetramethyl benzidine (TMB) produces a blue reaction product. The reaction is stopped with the addition of 2 M sulphuric acid which changes the colour to yellow; this is read at 450 nm. TMB is dissolved in DMSO (10 mg ml^{-1}) and this solution is slowly dispersed in 100 ml sodium acetate/citric acid buffer (0.1 M, pH 6.0) to give a final concentration of 0.1 mg ml^{-1}; 20 μl 30% hydrogen peroxide is added and the solution should then be used immediately.

    TMB is available commercially in an enhanced format, packaged as a two-part reagent (Cambridge Veterinary Sciences, Ely, UK).

2. 2,2'-Azinobis(3-ethylbenzthiazoline) sulphonic acid (ABTS) produces a green colour, and is read spectrophotometrically at 405 nm. ABTS is dissolved (0.5 mg ml^{-1}) in 0.1 M citrate/phosphate buffer pH 4.0 and 30% hydrogen peroxide added (10 μl/100 ml). The reaction can be stopped if desired with sodium fluoride solution (1.5%, w/v), although the reaction proceeds more slowly than with TMB.

A5.1.2   Alkaline phosphatase (AP) substrate

The usual substrate for AP is p-nitrophenyl phosphate (1 mg ml^{-1} in 0.1 M carbonate buffer pH 9.6 containing $MgCl_2$, 10 mM) which gives a yellow product absorbing at 405 nm. The reaction can be stopped by the addition of 3 M NaOH.

A5.2   Fluorometric substrates

The choice of enzyme/substrate systems here is either AP with 4-methyl umbelliferyl phosphate, or galactosidase with 4-methyl umbelliferyl-β-D-galactopyranoside. The phosphate derivative is dissolved (100 μg ml^{-1}) in the same buffer used for the colorimetric AP substrate, and the galactosidase substrate in 0.1 M phosphate buffer, pH 7.0.

Appendix 6:
ANTIBODY LABELLING METHODS

Labelling of antibodies should be carefully carried out because it is necessary to retain both the original binding properties of the antibody and the activity of the tracer molecule. The antibody should be at least partially purified by conventional salt fractionation or similar technique.

After conjugation, it is essential that all preparations are cleaned up by a chromatographic process such as gel filtration or FPLC to remove any polymerized products, unlabelled antibody or free tracer molecule. If stored at above 0°C a preservative may be necessary.

A6.1   Enzyme labelling

A6.1.1   Peroxidase label

A straightforward procedure involves the preparation of an aldehyde derivative of carbohydrate residues on horseradish peroxidase which is then coupled to antibody via amino groups (Wilson and Nakame, 1978). The following is a slight variation on the original method.

Purified antibody is dissolved in carbonate buffer (0.01 M, pH 9.6) and dialysed against the same buffer. Horseradish peroxidase (20 mg) is dissolved in distilled water (5 ml); sodium periodate ($NaIO_4$, 20 mg) is likewise dissolved (1 ml) and the two solutions mixed and stirred at room temperature for 20 min. The mixture is dialysed overnight at 1°C against 3 l acetate buffer (1 mM, pH 4.4). Sufficient carbonate buffer (0.1 M) is added to bring the solution to pH 9.5 and immediately antibody solution (50 mg protein) is added. The mixture is stirred at room temperature for 2 h before addition of $NaBH_4$ (0.5 ml, 4 mg ml^{-1}). The solution is stood at 4°C for 2 h and then dialysed overnight at 1°C against phosphate-buffered saline (as Appendix 7, A7.1 but with no Tween or Kathon). The preparation is clarified by separation and cleaned up. For long-term storage at $-70°C$ the authors have found that addition of 10 mg bovine serum albumin and 1 μl Kathon (see Appendix 7, A7.1) to 1 mg conjugate in 1 ml distilled water is suitable. The activity of the conjugate should be checked using a suitable substrate (see Appendix 5).

A6.1.2   Alkaline phosphatase label

A simple method for this label uses glutaraldehyde, which links proteins via amino residues (Engvall and Perlmann, 1972).

Purified antibody solution (0.1 ml, 5 mg protein/ml) is mixed with 1.5 mg alkaline phosphatase produced by centrifugation of a commercial suspension. The mixture is dialysed overnight at $+1°C$ against phosphate-buffered saline as above. Glutaraldehyde is added to the mixture to give a final concentration of 0.2% (v/v). After 2 h at room temperature, the mixture is diluted to 1 ml and again dialysed as above. The conjugate is cleaned up and stored at $4°C$ in 0.05 M Tris buffer, pH 8.0 (containing bovine serum albumin 50 g l^{-1}, $MgCl_2$ 0.001 M and NaN_3 0.2 g l^{-1}).

A6.2   Fluorochrome labelling

A thiourea bond is formed between the isothiocyanate derivative and the amino group of lysine in the antibody (Goding, 1983).

Purified antibody is dialysed against carbonate buffer (pH 9.5, Na_2CO_3 8.6 g l^{-1}, $NaHCO_3$ 17.2 g l^{-1}) to achieve optimal pH for conjugation. The fluorochrome (fluorescein isothiocyanate – FITC – or tetramethylrhodamine isothiocyanate – TRITC) is dissolved in DMSO and added dropwise to the stirred protein solutions; solution concentrations should be adjusted to give 10–15 μg fluorochrome/mg protein. Stirring should be continued for 2 h at room temperature and the cleaned-up conjugate can be stored at $4°C$, or at $-20°C$ in glycerol (50%, v/v) to prevent freezing. The absorbance of conjugates can be measured at the appropriate wavelength (495 nm for FITC or 550 nm for TRITC).

A6.3   Biotin labelling

The easiest method for coupling biotin to antibodies is to use N-hydroxysuccinimidobiotin; an amide bond is formed between biotin and lysine residues on the protein, releasing N-hydroxysuccinimide (Goding, 1983).

Purified antibody is dialysed against 0.1 M $NaHCO_3$, pH 8.0 and the protein concentration adjusted to 1.0 mg ml^{-1} with the same buffer. The biotin ester is dissolved in DMSO at the same concentration and added, with stirring, to the antibody solution at the rate of 120 μl ester/ml protein. A range of up to half or double this rate can be tried. Stirring should continue for 2 h at room temperature. After clean-up, the conjugate can be stored at $4°C$.

Appendix 7:
IMMUNOASSAY REAGENTS AND METHODS

A7.1 Buffers

For a general diluent (antibody or samples) and for washing plates, phosphate-buffered saline containing Tween (PBST) is used. This contains:

NaCl	8 g
KH_2PO_4	0.2 g
Na_2HPO_4	1.14 g
KCl	0.2 g
Tween 20	0.5 ml
Kathon	1.0 ml
Distilled water	1 litre
pH	7.4

Kathon is a proprietary preservative (Rohm and Haas Co., Philadelphia, USA). The authors find it preferable to thiomersal $(0.1 \text{ g } l^{-1})$.

For coating plates with antibody or protein antigens, a carbonate coating buffer is used. This contains:

Na_2CO_3	1.59 g
$NaHCO_3$	2.93 g
Distilled water	1 litre
pH	~9.6

For coating plates with bacterial cells, the optimum buffer is one containing methyl glyoxal:

Na_2HPO_4	13.4 g
$NaH_2PO_4.2H_2O$	0.8 g
Methyl glyoxal	3.0 ml
(0.3%, v/v aqueous solution)	
Distilled water	1 litre
pH	8.0

A7.2 Isotyping

To determine the class of a (monoclonal) antibody, kits are commercially available (e.g. Sigma Chemical Co.). Isotyping is necessary for several reasons, including the choice of an appropriate anti-species labelled antibody (see Section 5.7).

A7.3   Coating microtitration plates

Coating of plates with proteinaceous antigens (except bacterial cells) is done in the carbonate buffer described above. For determining the antibody titres in polyclonal sera or for screening hybridoma culture supernatants (see below), a coating concentration of 10 μg protein per ml will suffice. For competitive ELISAs, other (probably lower) concentrations may be optimal. The solution in buffer is applied to the plates (usually 200 μl/well) and these are then incubated for 2–3 h at 37°C. The plates are washed with five changes of distilled water, tapped on several layers of tissue to remove residual water and allowed to air dry. They can be stored with a desiccant at room temperature.

To coat with antibody for a sandwich assay, a suitable concentration is selected (if coating with a polyclonal, the authors routinely use a 1:1000 dilution of the serum initially) and diluted in coating buffer. Plate coating is then as described above, except that plates are washed with PBST rather than water, and stored in bags at −20°C.

For coating plates with bacterial cells, it is advantageous to use the buffer containing methyl glyoxal, which improves binding of cells to the plate. Otherwise, the procedure is as for other antigen-coated plates.

It is likely that after applying any of the above coating preparations, unoccupied binding sites will still remain on the surface of the microtitration plate wells. Even in the presence of PBST, there may therefore be some binding of second antibodies to those sites, increasing the assay blank (background) values. This can be prevented by application of a further coating buffer containing bovine serum albumin (10 g l^{-1}) to block those sites, after the primary coating has been applied. The blocking buffer should extend above the level in the wells which was used for the primary coating, e.g. 250 μl/well where the first coating was 200 μl/well.

A7.4   Example of protocol for a titration curve

Step 1   Prepare a dilution series of the serum in PBST, using tenfold steps. Add duplicate 200 μl volumes to wells of a suitable antigen-coated microtitration plate

Step 2   Incubate the plate, with a lid, for 1 h at 37°C
Step 3   Wash the plate × 5 with PBST
Step 4   Dilute appropriate enzyme-labelled anti-species antibody in PBST to the manufacturer's suggested working strength (often 1:1000). Add 200 μl/well
Step 5   Incubate the plate as before
Step 6   Wash plate as before
Step 7   Prepare substrate appropriate to the label used. Add 200 μl/well
Step 8   Observe colour development and read spectrophotometrically

A7.5   Example of protocol for limited reagent/competitive ELISA

Step 1   Prepare samples and standards, diluting as necessary in PBST. Add duplicate 100 μl volumes to wells on a suitable antigen-coated plate
Step 2   Dilute antibody preparation in PBST as required and add 100 μl to each well used in Step 1
Step 3   Incubate the plate, with a lid, for 1 h at 37°C
Step 4   Wash the plate × 5 with PBST
Step 5   Add anti-species antibody as above (see A7.4, Step 4)
Step 6   Incubate the plate as before
Step 7   Wash plate as before
Step 8   Finish assay with substrate (see A7.4, Steps 7 and 8).

Note: if the antibody in Step 2 is itself enzyme labelled, then Steps 5–7 are not needed.

A7.6   Example of protocol for excess reagent/sandwich ELISA

Step 1   Prepare samples and standards, diluting as necessary in PBST. Add duplicate 200 μl volumes to a plate coated with suitable capture antibody
Step 2   Incubate the plate, with a lid, for 1 h at 37°C
Step 3   Wash the plate × 5 with PBST
Step 4   Dilute the detector antibody in PBST as required and add 200 μl to each well
Step 5   Incubate the plate as before

Step 6   Wash the plate as before
Step 7   Add anti-species antibody as above (see A7.4, Step 4)
Step 8   Incubate the plate as before
Step 9   Wash the plate as before
Step 10   Finish assay with substrate (see A7.4, Steps 7 and 8)

Note: if the detector antibody (Step 4) is itself enzyme labelled, then Steps 7–9 are not needed.

Appendix 8:
PREPARATION OF IMMUNOAFFINITY COLUMNS

Suitable materials for preparation of immunoaffinity columns are available from Pharmacia LKB Biotechnology (Uppsala, Sweden), and the methods given below are taken from handbooks published by that company. If it is desired to elute live cells from the columns, then care should be taken to use buffer without preservative at that stage.

A8.1   Use of CNBr-activated adsorbent

CNBr-activated Sepharose 4B or 6MB is swollen with 1 mM HCl (200 ml/g gel) and washed in the same solution. Immunoglobulin is salted out from serum or monoclonal culture supernatant by standard protein chemistry methods and dissolved in a coupling buffer (0.1 M $NaHCO_3$, 0.5 M NaCl, pH 8.5). The optimum protein/gel ratio is about 5 mg Ig per ml gel and the amounts of IgG and gel used should be adjusted to give this loading. The gel is washed briefly with coupling buffer and transferred immediately to the antibody solution. After 2 h mixing at room temperature on an end-over-end mixer, the gel is washed (10 ml per ml gel) three times alternately in acetate buffer (0.1 M, 0.5 M NaCl, pH 4.5) and coupling buffer. If desired, unoccupied binding sites can then be blocked by using coupling buffer containing 1 M ethanolamine. Gels are packed into proprietary columns or in plastic syringes and storage should be in the cold with a preservative agent in the storage buffer.

A8.2   Use of protein-A-linked adsorbent

Protein A is isolated from *Staph. aureus* and binds to the Fc fragment of IgG from many animal species. Protein-A-linked Sepharose gels are supplied either dry, requiring swelling before use, or wet in a preservative solution; they must be washed in a suitable buffer before use. Columns can be prepared with antibody pre-bound to the gel; a better method, which uses less antibody, is to allow the antibody and analyte to bind in PBST and pass this preparation down the column.

A8.3   Elution of bound analyte from columns

Eluting agents based on several principles are available (Anon, 1987) and the choice will depend on the stability of the analyte in question and the method to be used to quantify it after desorption. It should be pointed out, however, that if it is desired to release live bacterial cells, then the toxicity of the eluting agent must be considered carefully.

Appendix 9:
SUPPLIERS OF EQUIPMENT AND CONSUMABLES

This is not an exhaustive list but is included as a guide to products available.

A9.1   Equipment

Pipettes: Anachem Ltd (Gilson pipettes)
Charles Street
Luton LU2 0EB
UK

ICN Flow
Eagle House
Peregrine Business Park
Gomm Road
High Wycombe
Bucks HP13 7DL
UK

Plate washers/readers:	ICN Flow
	Eagle House
	Peregrine Business Park
	Gomm Road
	High Wycombe
	Bucks HP13 7DL
	UK
	Dynatech Laboratories Ltd
	Daux Road
	Billingshurst
	West Sussex RH14 9SJ
	UK
	Skatron Ltd
	PO Box 34
	Studlands Park Avenue
	Newmarket
	Suffolk CB8 7DB
	UK
Data handling soft-	ICN Flow ('Titersoft')
ware:	Eagle House
	Peregrine Business Park
	Gomm Road
	High Wycombe
	Bucks HP13 7DL
	UK
	Beckman Instruments ('Immunofit')
	Progress Road
	Sands Industrial Estate
	High Wycombe HP12 4JL
	UK
	Pharmacia ('Multicalc')
	Wallac Oy
	PO Box 10
	SF-20 101 Turku
	Finland

Robotic equipment:	Beckman Instruments Progress Road Sands Industrial Estate High Wycombe HP12 4JL UK
Flow cytometers:	Chemunex S.A. 41 Rue de 11 Novembre 1918 94700 Maison-Alfort France
	Beckman Instruments ('FACScan') Progress Road Sands Industrial Estate High Wycombe HP12 4JL UK

A9.2   Consumables

Microtitration plates:	Gibco Life Technologies Ltd (Nunc plates) PO Box 35 Trident House Renfrew Road Paisley PA3 4EF UK
Labelled antibodies:	Sigma Chemical Co Ltd Fancy Road Poole Dorset BH17 7NH UK
	Pierce and Warriner (UK) Ltd 44 Upper Northgate Street Chester CH1 4EF UK
	Kirkegaard & Perry Laboratories Inc Dynatech Laboratories Ltd Daux Road Billingshurst West Sussex RH14 9SJ UK

Substrates and amplification systems:	Sigma Chemical Co Ltd Fancy Road Poole Dorset BH17 7NH UK
	Cambridge Veterinary Science Ltd Henry Crabb Road Littleport Ely Cambs CB6 1SE UK
	Pierce and Warriner (UK) Ltd 44 Upper Northgate Street Chester CH1 4EF UK
	Amersham International plc Amersham UK
	Kirkegaard and Perry Laboratories Inc Dynatech Laboratories Ltd Daux Road Billingshurst West Sussex RH14 9SJ UK
	IQ (Bio) Ltd Cambridge UK
Adjuvants:	Difco Laboratories Central Avenue East Molesey Surrey KT8 0SE UK
	Superfos Speciality Chemicals a/s Frydenlundsvej 30 Postboks 39 Dk – 2950 Vedbaek Denmark

Myeloma cell lines:	ICN Flow Eagle House Peregrine Business Park Gomm Road High Wycombe Bucks HP13 7DL UK
Affinity chromatography and particle separation:	Pharmacia Biosystems Ltd Biotechnology Division Davey Avenue Central Milton Keynes MK5 8PH UK
	Dynal (UK) Ltd Station House 26 Grove Street New Ferry Wirral Merseyside L62 5AZ UK
	Pierce and Warriner (UK) Ltd 44 Upper Northgate Street Chester CH1 3EF UK
Bacterial cultures:	National Collection of Type Cultures Central Public Health Laboratory Colindale Avenue London NW9 5HT UK
	American Type Culture Collection 12301 Parklawn Drive Rockville Maryland 20852–1776 USA

Hybridoma culture media:	Gibco Life Technologies Ltd PO Box 35 Trident House Renfrew Road Paisley PA3 4EF UK
Plasticware for hybridoma culture:	Northumbria Biologicals Ltd (Costar Europe Ltd) South Nelson Industrial Estate Cramlington Northumberland NE23 9HL UK
	ICN Flow Eagle House Peregrine Business Park Gomm Road High Wycombe Bucks HP13 7DL UK
Immunoassay kits:	bioMerieux Ltd ('Tecra' assays) Grafton House Grafton Way Basingstoke Hampshire RG22 6HY UK
	Organon Teknika Ltd Science Park Milton Road Cambridge CB4 4FL UK
	Oxoid Unipath Ltd Wade Road Basingstoke Hampshire RG24 0PW UK

Cortecs Diagnostics Ltd
Techbase 1
Newtech Square
Deeside Industrial Park
Deeside
Clwyd CH5 2NT
UK

Rhone-Poulenc Diagnostics Ltd
Montrose House
187 George Street
Glasgow G1 1YT
UK

References

Adams, A. N. and Barbara, D. J. (1982) The use of Fab-based ELISA to detect serological relationships among carlaviruses. *Ann. Appl. Biol.* 101, 495–500

Anon (1975) Fluorescent antibody method official first action. *J. Assoc. Off. Anal. Chem.* 58, 417–419

Anon (1987) *Affinity Chromatography, Principles and Methods*, pp. 92–94, Pharmacia LKB Biotechnology, Uppsala, Sweden

Barrell, R. A. E. and Paton, A. M. (1982) The detection of salmonellas in foods and environmental samples by a semi-automatic method. In: *Isolation and Identification Methods for Food-poisoning Organisms* (Eds J. E. L. Corry, D. Roberts and F. A. Skinner), Society for Applied Bacteriology Technical Series no. 17, pp. 77–81, Academic Press, London

Beckers, H. J., van Leusden, F. M., Meijssen, M. J. M. and Kampelmacher, E. H. (1985) Reference material for the evaluation of a standard method for the detection of salmonellas in foods and feeding stuffs. *J. Appl. Bact.* 59, 507–512

Bird, J. A., Easter, M. C., Hadfield, S. G., May, E. and Stringer, M. F. (1989) Rapid *Salmonella* detection by a combination of conductance and immunological techniques. In: *Rapid Microbiological Methods for Foods, Beverages and Pharmaceuticals*, Society for Applied Bacteriology Technical Series no. 25, pp. 165–183, Academic Press, London

Brook, M. G. and Bannister, B. A. (1991) Verocytoxin producing *Escherichia coli*. *Br. Med. J.* 303, 800–801

Butler, V. P. and Chen, J. P. (1967) Digoxin-specific antibodies. *Proc. Natl Acad. Sci.* 57, 71–78

Dodd, C. E. R. (1990) Detection of sites of microbial growth in food by cryosectioning and light microscopy. *Food Sci. Tech. Today* 4, 180–182

Ekins, R. P. (1983) The precision profile: its use in assay design, assessment and quality control. In: *Immunoassays for Clinical Chemistry*, 2nd ed (Eds W. M.

Hunter and J. E. T. Corrie), pp. 76–104, Churchill Livingstone, Edinburgh

Engvall, E. and Perlmann, P. (1971) Enzyme-linked immunosorbent assay (ELISA). Quantitative assay of immunoglobulin G. *Immunochemistry* 8, 871–874

Engvall, E. and Perlmann, P. (1972) Enzyme-linked immunosorbent assay III. Quantitation of specific antibodies by enzyme-labelled anti-immunoglobulin in antigen-coated tubes. *J. Immunol.* 109, 129–135

Firstenberg-Eden, R. (1986) Electrical impedance for determining microbial quality of foods. In: *Foodborne Microorganisms and Their Toxins: Developing Methodology* (Eds M. D. Pierson and N. J. Stern), pp. 129–144, Marcel Dekker, New York

Flowers, R. S. and Klatt, M. J. (1989) Immunodiffusion screening method for detection of motile *Salmonella* in foods: collaborative study. *J. Assoc. Off. Anal. Chem.* 72, 303–311

Flowers, R. S., Klatt, M. J. and Keelan, S. L. (1988) Visual immunoassay for detection of *Salmonella* in foods: Collaborative Study. *J. Assoc. Off. Anal. Chem.* 71, 973–980

Goding, J. W. (1983) *Monoclonal Antibodies: Principles and Practice*, pp. 223–230, Academic Press, London

HMSO (1990) *The Microbiological Safety of Food*, part I (Chairman Sir M. Richmond), Her Majesty's Stationery Office, London

HMSO (1991) *The Microbiological Safety of Food*, part II (Chairman Sir M. Richmond), Her Majesty's Stationery Office, London

Huse, W. D., Sastry, I., Iverson, S. A., Kang, A. S., Alting-Mese, M., Burton, D. R., Benkovic, S. J. and Lerner, R. A. (1989) Generation of a large combinatorial library of the immunoglobulin repertoire in phage lamda. *Science* 246, 1275–1281

Ibrahim, G. F., Fleet, G. H., Lyons, M. J. and Walker, R. A. (1985) Method for the isolation of highly purified Salmonella flagellins. *J. Clin. Micro.* 22, 1040–1044

Kohler, G. and Milstein, C. (1975) Continuous culture of fused cells secreting antibody of defined specificity. *Nature* 256, 495–497

Langman, R. E. (1972) The occurrence of antigenic determinants common to flagella of different *Salmonella* strains. *Eur. J. Immunol.* 2, 582–586

Lee, H. A., Wyatt, G. M., Bramham, S. and Morgan, M. R. A. (1989) Rapid enzyme-linked immunosorbent assays for the detection of *Salmonella enteritidis* in eggs. *Food Agric. Immunol.* 1, 89–99

Lee, H. A., Wyatt, G. M., Bramham, S. and Morgan, M. R. A. (1990) Enzyme-linked immunosorbent assays for *Salmonella typhimurium* in food: feasibility of 1-day *Salmonella* detection. *Appl. Environ. Microbiol.* 56, 1541–1546

Lee, H. A. and Morgan, M. R. A. (1990) Detection of microbial toxins by immunological techniques. In: *Development and Application of Immunoassay for Food Analysis* (Ed. J. H. Rittenberg), pp. 171–200, Elsevier Applied Science, London

McBroom, C. R., Samanen, C. H. and Goldstein, I. J. (1972) Carbohydrate antigens: coupling of carbohydrate to proteins via diazonium and phenylisothiocyanate reactions. In: *Methods in Enzymology*, vol. XXVIIB, pp. 212–219, Academic Press, New York

Mirhabibollahi, B., Brooks, J. L. and Kroll, R. G. (1990) An improved amperometric immunosensor for the detection and enumeration of Protein A-bearing *Staphylococcus aureus*. *Lett. Appl. Microbiol.* 11, 119–122

Newell, D. G., McBride, H. and Pearson, A. D. (1984) The identification of outer membrane proteins and flagella of *Campylobacter jejuni*. *J. Gen. Microbiol.* 130, 1201–1208

Notermans, S. and Werners, K. (1991) Immunological methods for detection of foodborne pathogens and their toxins. *Int. J. Food Microbiol.* 12, 91–102

Oste, C. (1988) Polymerase chain reaction. *Biotechniques* 6, 162, 167

Pettipher, G. L. and Rodrigues, U. M. (1982) Rapid enumeration of microorganisms in foods by the direct epifluorescence filter technique. *Appl. Environ. Microbiol.* 44, 809–813

Pinder, A. C., Purdy, P. W., Poulter, S. A. G. and Clark, D. C. (1990) Validation of flow cytometry for rapid enumeration of bacterial concentrations in pure cultures. *J. Appl. Bact.* 69, 92–100

PHLS (1990) *Communicable Disease Report*, Public Health Laboratory Service, Communicable Disease Surveillance Centre, London

Sharpe, A. N. (1980) *Food Microbiology – A Framework for the Future*, C. C. Thomas, Springfield, Illinois, USA

Smith, R. W. and Koffler, H. (1971) Production and isolation of flagella, In: *Methods in Microbiology*, Vol. 5A (Eds J. R. Norris and D. W. Ribbons), pp. 165–172, Academic Press, London

Stanley, P. E. (1989) A concise beginner's guide to rapid microbiology using adenosine triphosphate (ATP) and luminescence, In: *ATP luminescence*, Society for Applied Bacteriology Technical Series no. 26, pp. 1–10, Academic Press, London

Stewart, G. S. A. B. (1990) *In vivo* bioluminescence: new potentials for microbiology. *Lett. Appl. Microbiol.* 10, 1–8

Sutherland, I. W. and Wilkinson, J. F. (1971) Chemical extraction methods of microbial cells. In: *Methods in Microbiology*, Vol. 5B (Eds J. R. Norris and D. W. Ribbons), pp. 346–383, Academic Press, London

van Weeman, B. K. and Schuurs, A. H. W. M. (1971) Immunoassay using antigen–enzyme conjugates. *FEBS Lett.* 15, 232–236

Ward, C. M., Chan, H. W-S. and Morgan, M. R. A. (1988) The potential of fluorescence detection in ELISA. In: *Immunoassays for Veterinary and Food Analysis-1* (Eds B. A. Morris, M. N. Clifford and R. Jackman), pp. 275–278, Elsevier Applied Science, London

Ward, E. S., Gussow, D., Griffiths, A. D., Jones, P. T. and Winter, G. (1989) Binding activities of a repertoire of single immunoglobulin variable domain secreted from *Escherichia coli*. *Nature* 341, 544–546

Weeks, I., Beheshti, I., McCapra, F., Campbell, A. K. and Woodhead, J. S. (1983) Acridinium esters as high specific activity labels in immunoassay. *Clin. Chem.* 29, 1474–1479

Wilson, M. B. and Nakane, P. K. (1978) Recent developments in the periodate method of conjugating horseradish peroxidase to antibodies. In: *Immunofluorescence and Related Staining Techniques* (Eds W. Knapp, K. Holubar and G. Wick), pp. 215–224, Elsevier/North Holland Biomedical Press, Amsterdam

Wilkinson, A. P., Ward, C. M. and Morgan, M. R. A. (1991) Immunological analysis of mycotoxins. In: *Modern Methods of Plant Analysis*, vol. 13 (Eds H. F. Linskens and J. F. Jackson), in press, Springer-Verlag, Berlin

Wyatt, G. M., Langley, M. N., Lee, H. A. and Morgan, M. R. A. (1991) Salmonella immunoassay or immunoenrichment – a chicken and egg situation? In: *Food Safety and Quality Assurance: Applications of Immunoassay Systems* (Eds C. J. Smith and M. R. A. Morgan), in press, Elsevier Applied Science, London

Yalow, R. S. and Berson, S. A. (1959) Assay of plasma insulin in human subjects by immunological methods. *Nature* 184, 1648

Index